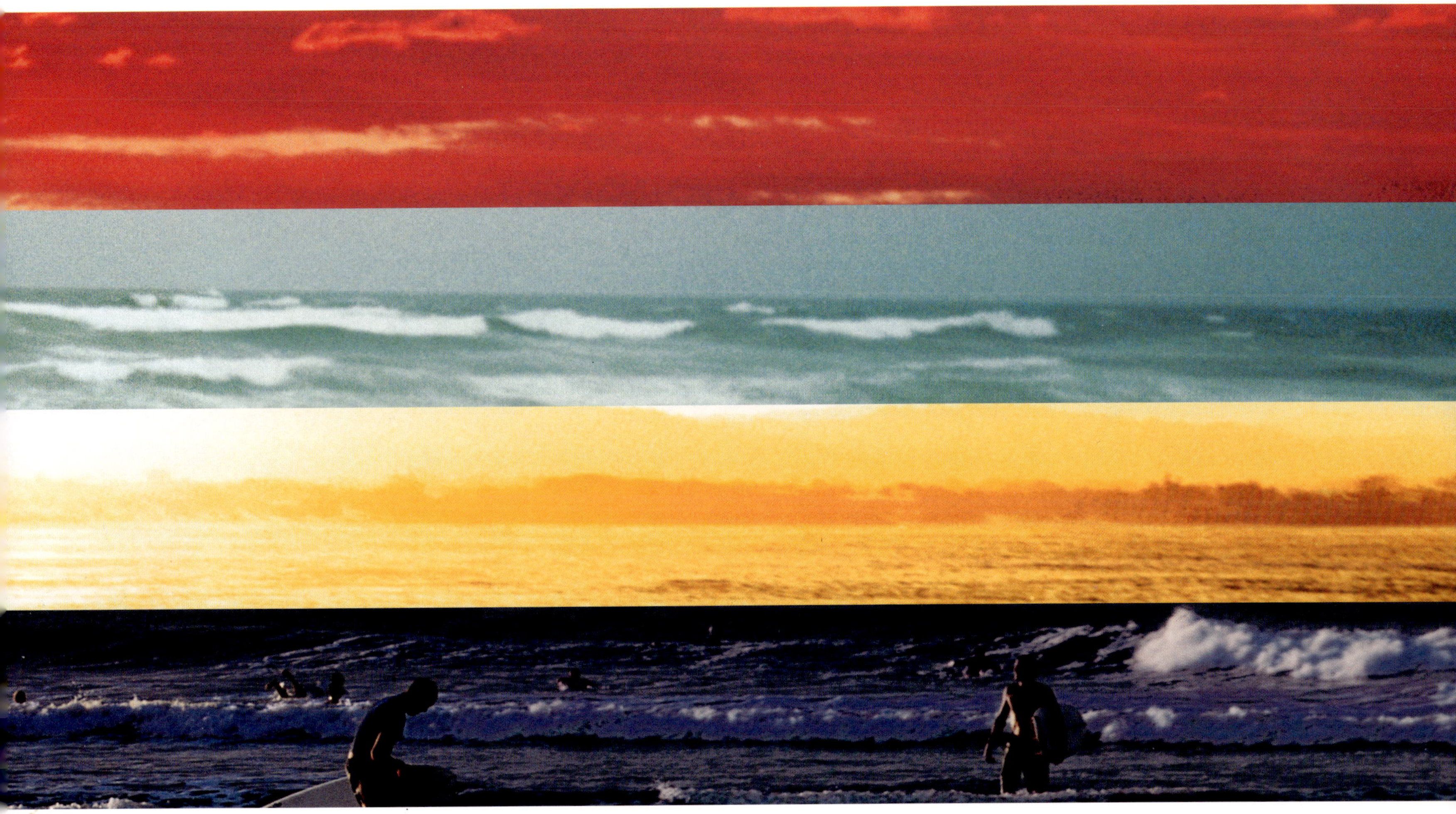

SURF BOOK

Photographs by Michael Halsband
Text by Joel Tudor

The Surf Book featuring:

Bud Browne, The Irons Brothers, Dale Velzy, John Peck, Wayne Lynch, Dick Brewer, Matt Howard & Britt Quinn, George Downing, David Nuuhiwa, Mike Hynson, Rob Machado, Donald Takayama, The Keaulanas, Dorian Paskowitz, Mark Cunningham, LeRoy Grannis, Hoole/McCoy, Skip Frye, Thomas Campbell, Duncan Campbell, Bob McTavish, Woody Brown, Jock Sutherland, Robbie Page, Beau Young, Billy Hamilton, The Fletchers, CR Stecyk, Mike Diffenderfer, Tony Alva, Mark Martinson, George Greenough, Tony Caramanico, Kelly Slater, Strider Wasilewski, J Brother, Tom Morey, Nat Young, Tom Curren, Art Brewer, Gerry Lopez, Donovan Frankenreiter, Miki Dora, and Hawaii.

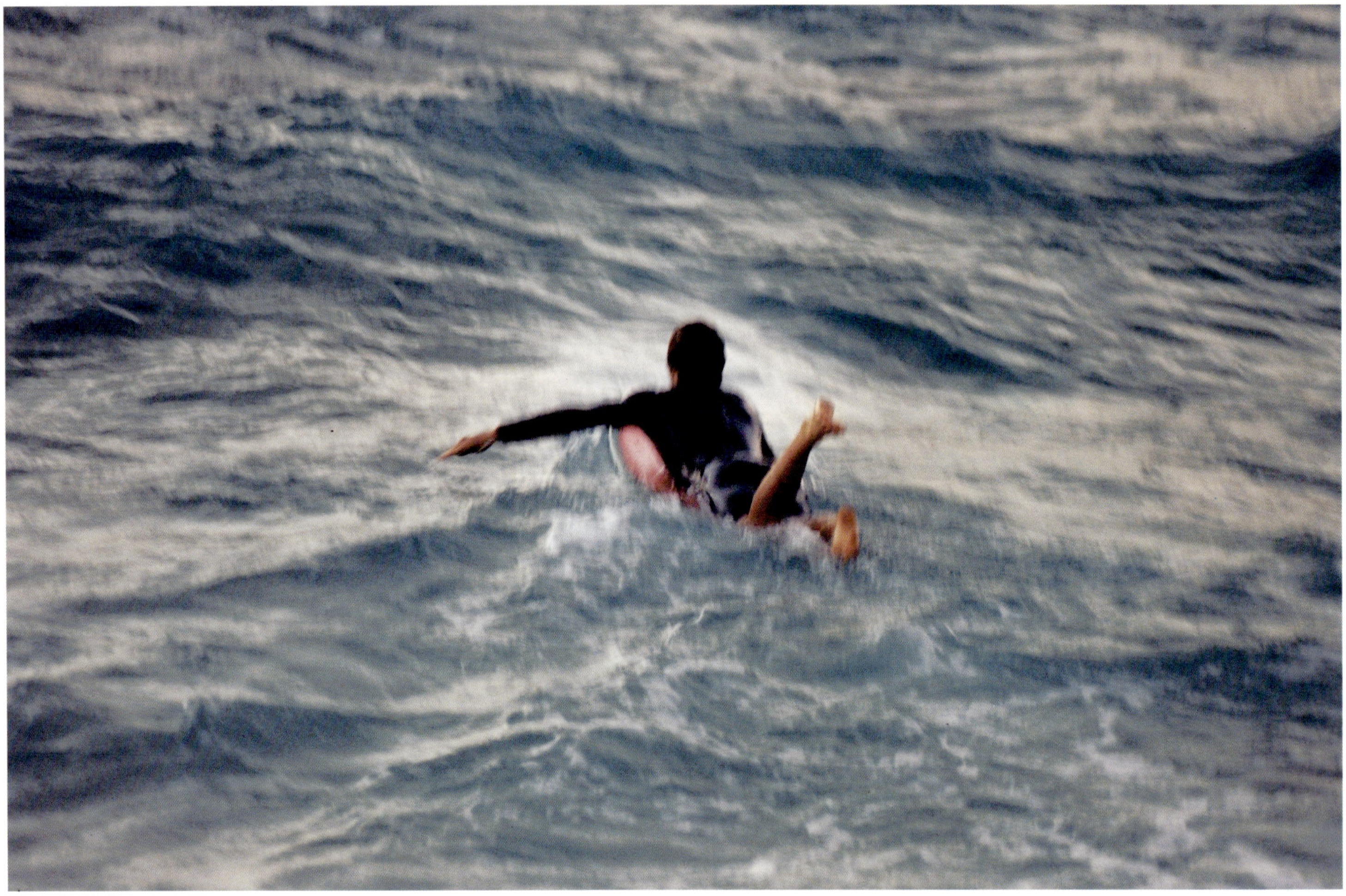

Preface

I grew up in Manhattan, and for as long as I can remember I dreamt of being a surfer. I got my first taste when I was twelve years old, when my family went on vacation to South America. After years of coming close, I finally decided to start surfing full on. Little did I know that one day being a professional photographer would help bring me into the heart of surf culture and history. During the past four years, I have been accepted into this microcosm in a way that makes me feel as if I have been surfing my whole life. The person that made this possible is the legendary Joel Tudor. The adventure I have been on with him has exposed me to people who have been crucial to the evolving technology and culture of surfing. I have sat, listened to, and photographed these individuals as they opened up with Joel in ways they never would have with anyone else. There is a mutual respect between Joel and these individuals that is both deep and loving.

This book is a record of a dream come true, the scope of which I couldn't have imagined until now. I have been given the opportunity to see the world of surfing through Joel's eyes. As someone on an endless spiritual quest, he has traveled the entire planet, met almost everyone who is part of the surf culture, and has a perspective and breadth of knowledge like no one else. This book is a slice of that world—a portrait of Joel through his influences. His words are evidence of what he has learned and the conclusions he has drawn from this culture.

Years before Tony Caramanico introduced me to Joel in the Ditch Plains parking lot in Montauk, New York, I read about Joel in surf related magazines and saw him in surf films. He was the living embodiment of what surfing meant to me. When people talked about "soul surfing," it was Joel's approach that defined that term for me. From growing up an hour from the ocean, near the town *Fast Times at Ridgemont High* was based on, to living the surfer's ultimate dream life, he is still as excited as if it were his first day. He radiates a boundless love for surfing. The people he attracts are from both ends of the spectrum—the oldest living legends to those in their first thrill.

I never intended to mix my photography with surfing. I did think that if I were ever to photograph or film surfing, Joel would be the one person I would want to collaborate with. Yet I saw no way for us to ever meet or even connect. I could never have predicted the scale of what unfolded, but when Joel approached me about working on a project together, I didn't stop to think about it. I immediately said yes, and we set a date right there to start the following week at his house in Southern California. I knew that if I were ever going to learn more about the history of surfing culture and technology Joel would be the best and only guide I would want. I could not predict I would spend the next four years learning from the people who created and defined surfing culture and technical knowledge. Living at Joel's house and meeting the people who passed through it, sleeping in the attic of Donald Takayama's Hawaiian Pro Design factory, and traveling to Hawaii and Australia with Joel and Donald—all these experiences helped me view and be personally subjected to the surfing world through their eyes. Spending all this time with different members of this widespread community had a profound and magical effect on me. The fact that I was risking the commercial photography business I had built in New York was secondary. I was on a big adventure, and I didn't want to stop to figure out how I was going to finance it or support myself. I believed everything would work itself out. This was the first thing I learned from surfing—let go of your worries and everything will work out. *Thanks Izzy.* Beyond how much I learned from this project, it has enriched and therefore changed the way I live, and the people documented in this book are the ones I am happy to thank.

Photography has been my passport to any world that I have been interested in—fashion, ballet, rock music, New York art and the art world, and surfing. I always enter one of these worlds as a photographer, which involves me as an insider. For me, photography is the only way to register the true subjective experience of the cultures that I have been interested in. In this project, my experience was very different, without Joel I would never have been able to get as close to these people as I did. People respect and admire Joel not just for his surfing but for what he has stood for and believed in, despite the criticisms he has endured for choosing a very different path from that of most professional surfers. I would like to think this is something Joel and I have in common.

Conceptual photographers want to control their world by making all the decisions and setting up the scenario. The artificiality of this approach has always bothered me. I felt that if I were able to experience the situation in as real a way as possible and make the picture at the same moment, that energy would be documented. I am always trying to combine the technical control and formal approach of studio photography with the "shoot from the hip" looseness of street photography or the "run and gun" of documentary films.

Photography is what first grabbed me. I wasn't into the image of the photographer and how the photographer was defined. Those things have to do with a style of living. In art school, I found that I could define photography for myself. I used criticism to open my eyes further and to explore other possibilities. If you love doing something, it doesn't seem like work, and you look for more of what makes you love it so much. Despite all the setbacks, disappointment, and aggravation, the love of something is what keeps you returning to it. What both Joel and I have chosen to do in our lives is ultimately lonely. We may be with other people when we do it, but we are always alone in the process. This isolation is what all of us have in common. Perhaps we can learn to respect that in ourselves and in each other.

I hope you enjoy this book as much as I enjoyed making it.

– Michael Halsband

Introduction

It's dusk in Manhattan, but the only indication that we're in NYC is a signature of thin light through the scrimmed windows of the studio.

We're in our twelfth hour of photo editing here at Michael Halsband's clean, well-sited studio near Herald Square. Joel's meal—messengered over from one of only two restaurants he approves of—is stinking up the joint. Soy gluten medallions swim in some sort of garlic swamp, and combined with the reek of the chestnut vendor downstairs it's all a bit heady. A bar band from downtown caterwauls on the hi-fi. The studio's wooden floors are varnished to a Bristol glow, so laps on the skateboard are stress relievers. If we're good soldiers and get our work done by nightfall, it's off to Montauk to meet a supposed swell.

Ah, but Michael Halsband—dubbed "Japanese Finger" by George Greenough—has accumulated a stack of proofs, contact sheets, slides, and prints comparable in scale to the accumulated chattel of William Randolph Hearst. It's our duty to weed these down into a respectable English garden. We repeatedly tell Halsband that some of his babies are absolute ogres . . . always a touchy moment in the ego-bruised world of "creatives." It's at this instant that the magic of the project coalesces. "Great," he smiles. "We really need to cut through this."

No hemming and hawing, no anguished faces. I give Tudor a "Hey, we've got a pro on our hands here" sort of glance. We all bust up laughing for different reasons. Personally, I'm howling over the irony of us philistines calling the artistic shots based on some half-baked surf quotient.

Michael Halsband's apparent lack of armor plating is the key to his success, and the reason he and Joel have been able to round up what can best be described as the finest archive of living surf maestros in one file.

How'd it all start?

A young surfer from California grew up in the company of living legends, and he turned down the squelch of youth enough to listen. Each year, the worth of their lessons—both overt and unspoken—grew in value. Throughout his development he continued to interface with older surfers and to solidify his skills on a variety of surfboards, most notably at

Pipeline. The fact that the collective wisdom of surfing's elders was available unfiltered wasn't lost on him. From the age of thirteen he made it a point to soak up the words of past champions, master board builders, and underground cult figures.

This kid—this Joel Tudor—found more grist in the style of the elders than he did in the contemporary surfscape, and it showed in the way he rode. By eighteen, his was the appellation of default when it came to classical wave-riding virtuosity. When he sank a rail at Pipe, it referenced Butch Van Artsdalen and Gerry Lopez. When he walked his board, the obvious influence of Nat Young came to the fore; after all, it was Nat who told him that the rider should "walk in place," creating the illusion of the board sliding effortlessly beneath him. And blow me down if that isn't a hit of Wayne Lynch in his work on double-ended eggs. World acclaim—and titles—ensued. He has since become one of the most relentlessly documented surfers expelling warm air.

Realizing how much he'd received, the young surfer wanted to give thanks in the best way he knew. He culled a short list of august surf photographers who might deign to chronicle his personal pantheon. The time and investment required called for an entrepreneurial individual with a rather exotic skill set. Tudor spends a fair piece of time in New York, drawn to the art storm of Gotham as much as the waves on the East End. To serve both, he keeps a quiver in Montauk and a bicycle and bedroll in Greenwich Village. When Joel was introduced to a soaking-wet photographer from Manhattan at the Ditch Plains parking lot in 1999, he suspected he'd found his man.

Halsband is a bit of a legend himself, though not in the surfing world. Upon graduating from the School of Visual Arts in 1980, he was promptly scouted by Mick Jagger to document the Rolling Stones' Tattoo You tour in '81. He became a friend and confidante of both Jagger and Keith Richards, who counted on him to relay the vibe of their program. Prior to that he had worked closely with Andy Warhol and his young protégé Jean-Michel Basquiat. His photography for their 1985 two-man show became a touchstone of postwar American pop culture.

Throughout the following decade, Michael focused on fashion and portrait assignments in art, music, and entertainment. Earning a reputation as a trustworthy and artful photographer who relentlessly "got the job done." Subjects included Francis Ford Coppola, Kim Cattrall, Hunter S. Thompson, Johnny Depp, Louise Nevelson, Keith Haring, Federico Fellini, Curtis Blow, Iggy Pop, Beck, Lucinda Williams, Al Pacino, David Bowie, and The Ramones. As an adjunct to the competitive gruel, he assigned himself a long-term project documenting The School of American Ballet. For six years he employed a strict available-light discipline that yielded a stunning-by-any-measure suite of stark, honest images of ballerinas sweating it. While satisfied with the results, it was the process that proved the biggest win, and the opportunity to explore new techniques, equipment, and aesthetics that led him to his next personal assignment: a long look under the lid of the sex industry.

"It was originally about hardcore pinups," he says. "But after getting to know the people, I couldn't objectify them in that manner. That scene is a black hole of confusion."

Ironically, much like surfing. "This surf journey has been life changing. If I put it in some hierarchy, I'd say without hesitation that it's been as interesting as touring with the Stones." What started as a hobby away from the city transited into a three-year tour girdling the globe.

"Joel, of course, was the catalyst for this," he says. "It's been amazing, but it's tested me as well. In many ways, Joel is like Basquiat. Jean-Michel couldn't really function in the world; he made no plans and rarely adhered to a schedule of any sort. They both have a magic combination of childlike innocence and intuitive ability—what I mean is that they both predict what you want and try to give it to you . . . and like Basquiat—though for different reasons—Joel spends a lot of time and energy trying to escape from the world through various channels."

The modus of the project was quite the opposite. It was about embracing the world and its wave-riding habitués in a direct, person-to-person way.

It started at WindanSea, shooting Australian mensch/guru Wayne Lynch. In a way, Joel was using this project as a test. Surfers use peculiar radar to vet out the creeps on dry land. Lynch and Halsband hooked right up, and the show was on.

"Michael gets along with everybody, because he's sincere and he's a listener," Joel says. "He doesn't come on hard, and he's obviously really into surfing and its myths and heroes. But that doesn't mean he doesn't get it done. I saw him hammer more photos of guys who previously hated photographers than I ever would have imagined."

If you're a surfer, you know that names like Miki Dora, Mike Hynson, Barry Kanaiaupuni, and George Downing are freighted with the sort of specific density that makes plutonium look like a dessert topping. These gentlemen don't suffer fools. If you front, you WILL get shut down. At best. Most likely, you won't make it to "hello." They're used to fielding a phone call each year that goes something like, 'Hi, this is Johnny Rocket from *Surf* magazine, and

I'd like to ask you about the current state of (insert cliché topic here). Click. Bzzzz. "If you'd like to make a call, please hang up and . . ."

Joel's reputation among the old guard was invaluable here. Put simply—and literally—they like his style. The louder, more garish, more megalomanic, more commoditized surfing gets, the more Joel's minimalist values stand out. When everyone's screaming, it's often the whisper that's heard. And in a time where wave riding's senses are swept under the carpets, the significance of a kid paying tribute doesn't go unnoticed.

Yet this only got the project so far.

Halsband had to get past the metal detectors and into the sanctum. That's where his warmth and sincerity came into play. He is who he is, and what you see is most certainly what you get. You drop your guard, you share a joke, and before you know it, he's changed film five times, and your soul is being Fedex'd to a lab in NYC.

Humble to the hilt, Michael was mainly concerned that Tudor's comments on the subjects would come through loud and clear. It's not that he wanted his photos to play a supporting role—this is, after all, a book of portraits—it's that the theme of influence could only really be told through Joel's vignettes. He had bathed in the glow of Joel's road stories on long international flights—tales of the masters retooled through the kid's vigorous phrasing. Seeing transcriptions of those same stories stripped of their breakdowns and false starts, rendered to their fat-free essence, struck Halsband as denuded and altogether too short. We posited that Michael's portraits were loaded with information and pathos, that he'd captured each participant in a way that was suitably deep, and the Joel word-bursts anchored them to the page, giving further context.

Identifying the book's audience proved vague. Generally, it's for those interested in surfing. The theme itself is clearer: knowledge is on tap for the young surfer who might only seek and listen. It was decided that biographical information on each subject wouldn't have a place here. That's already been laid down. If you need to be told that Mike Hynson starred in *Endless Summer*, that John Peck is nontraditional, and that Dick Brewer is the key holder, well, sit back and just enjoy the pictures. Perhaps they'll inspire you to delve deeper.

The selection of the surfers themselves was a subjective exercise based on people Joel felt had an effect on his life. They range from close, dear friends like Donald Takayama and Nat Young to fellows he barely knows, like Billy Hamilton and Jock Sutherland. Also included are a handful of artists and filmmakers that Tudor has worked with, been inspired by, and trusts: Craig Stecyk, Art Brewer, J Brother, and Thomas Campbell. With a single exception, all who were asked to participate in this book responded in the positive. Some who figure in Joel's sphere of influence aren't represented here, despite his best intentions. And while this is a relatively watershed assemblage of surfing's greats, trust us when we point out that it's only a surface scratch.

The photographic process led from New York to Europe; from La Jolla to the North Shore; from Malibu to Oregon; and from Noosa Heads to WindanSea. From the picnic bench at Makaha to the storage room at ninety-year-old Bud Browne's house in Costa Mesa, Tudor and Halsband hauled their kit to every surfer on the list. And of course, the list grew as time passed. "Wait a minute," Joel might say. "Before I surfed, I skated, and Jay Adams and Tony Alva surfed and skated and influenced me . . . I wonder if we can chase them down?" Michael would build a hole into his schedule, and off they'd go.

Joel says, "It's really crucial that readers see this for what it is. We're all working on ghetto rate here, from me to the publisher to Michael to the writer to the designer. It's a small investment we're all making in helping give these guys the respect they deserve, and to say thank you for laying down the template for the life we all benefit from."

Which brings us back to Broadway and Twenty-ninth, where the closest visual H2O references are the wooden water tanks on top of every building surrounding us. Eyes and nerves are raw, and mild dehydration is setting in. We've made it through our first edit, and we're gathered around the table looking at a single image. It's Bud Browne, rail thin on his front porch, thanking you for coming by and entering his surfing life.

– Scott Hulet

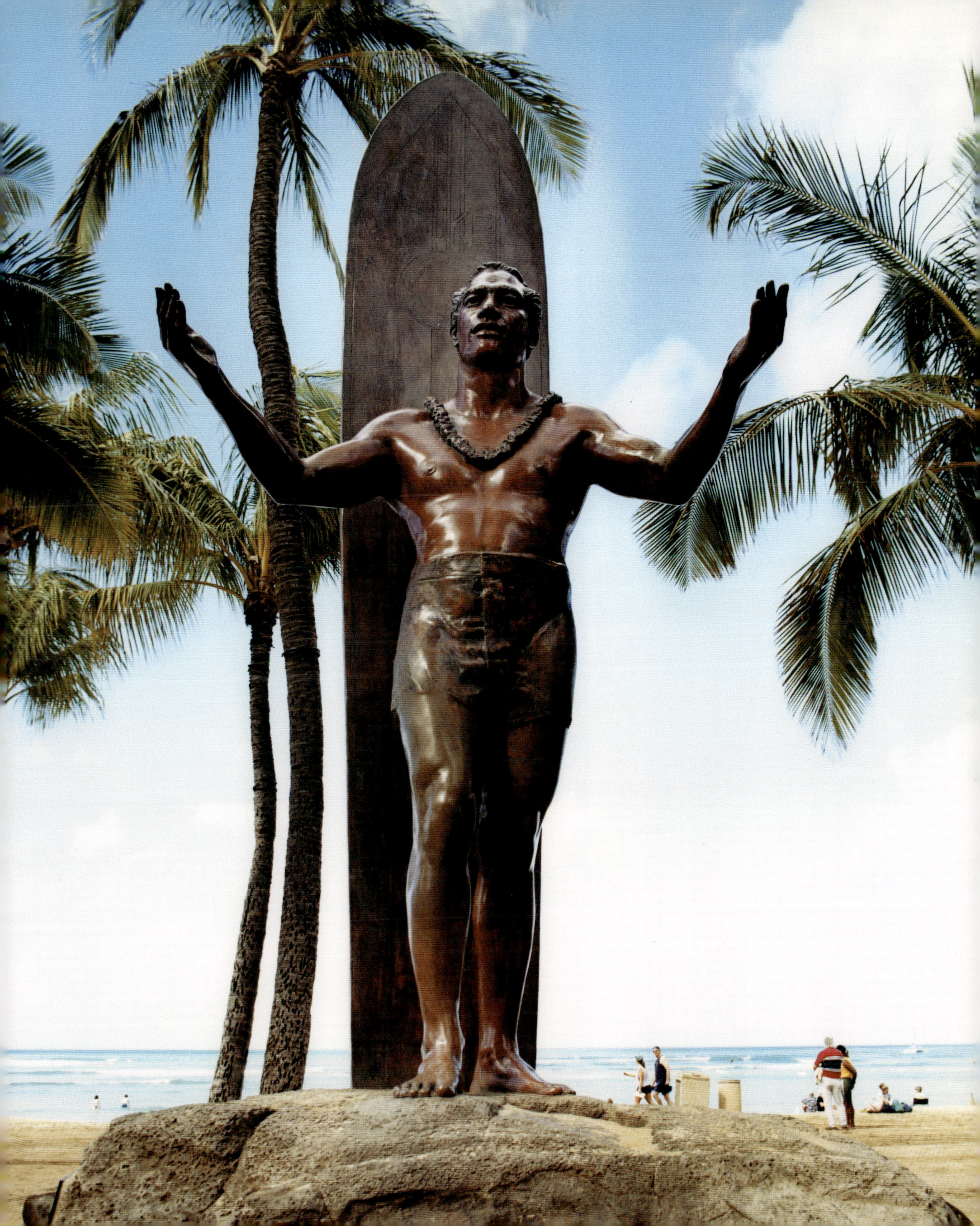

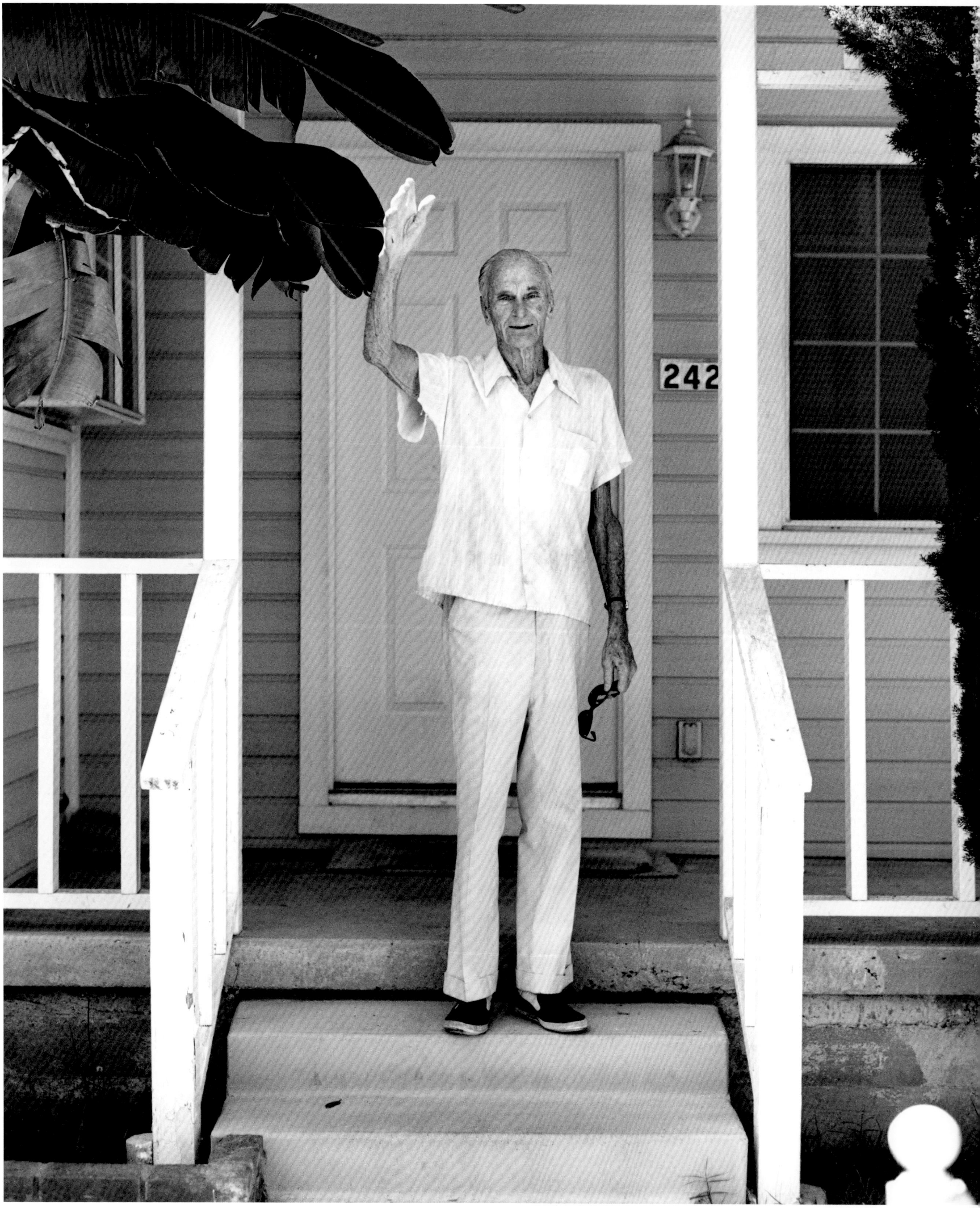
242

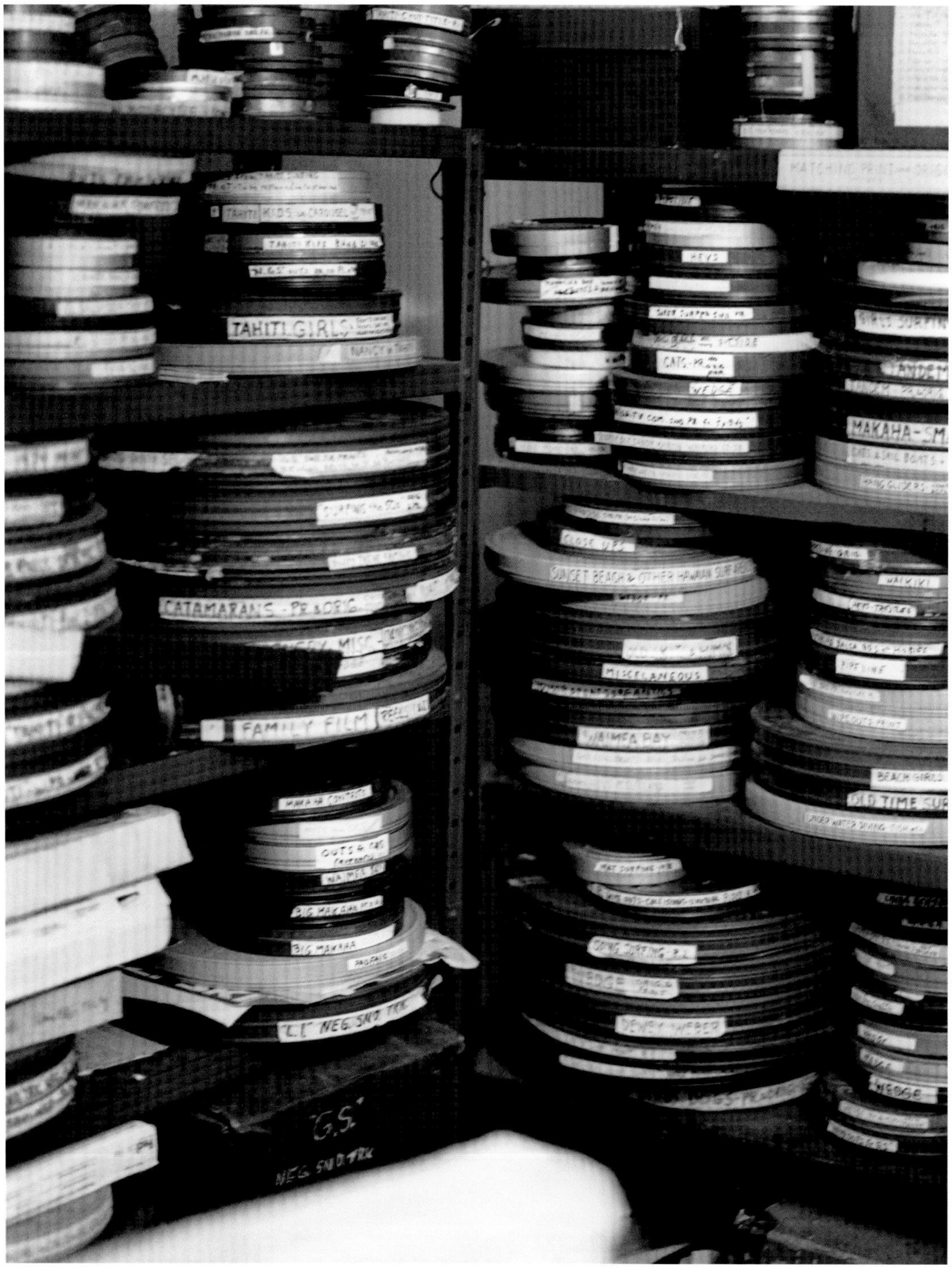
TAHITI GIRLS
CATAMARANS - PR & ORIG.
FAMILY FILM
MISCELANEOUS
WAIMEA BAY
BIG MAKAHA
"L.I." NEG. SND TRK
BEACH GIRLS
"G.S."
NEG. SND TRK

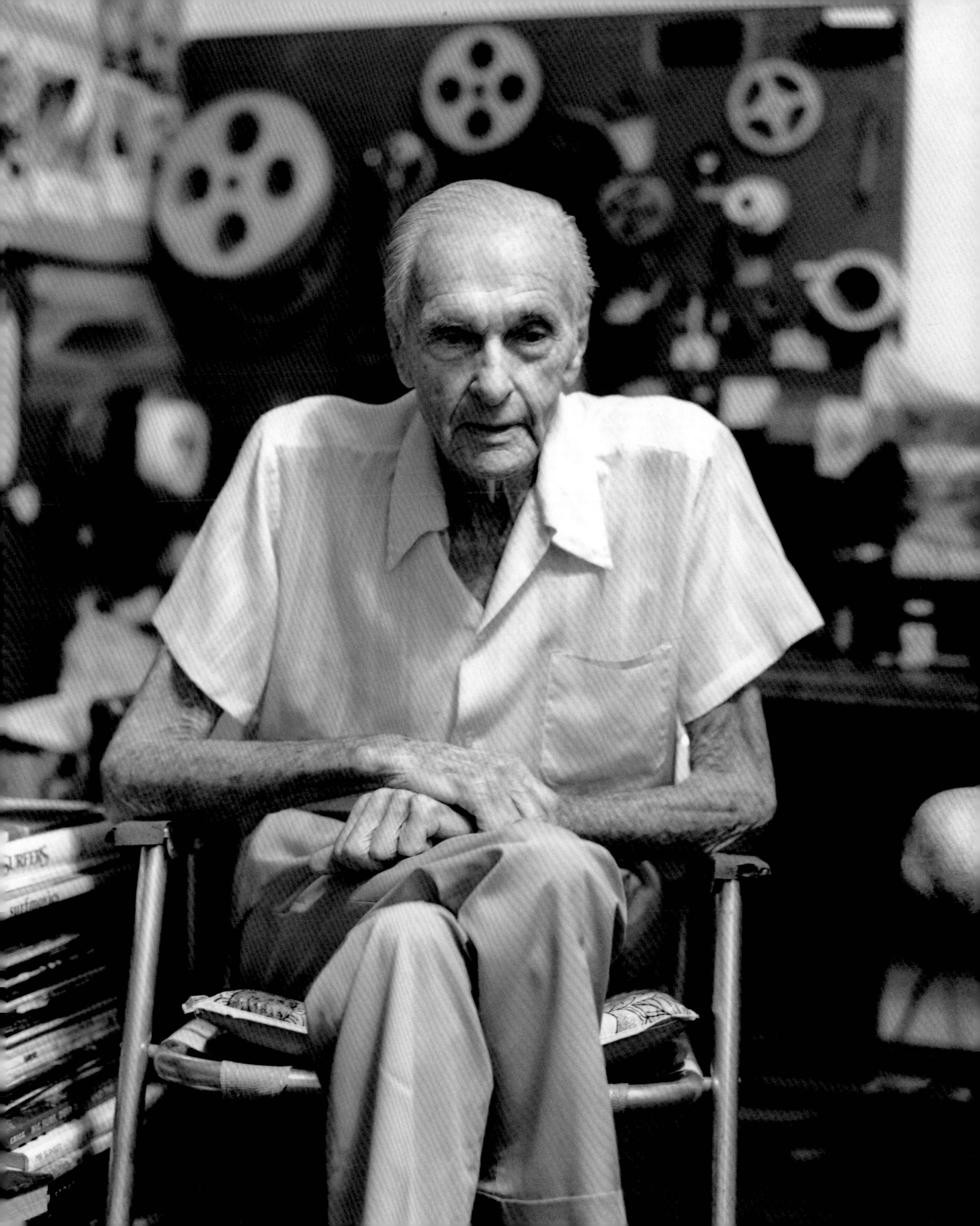

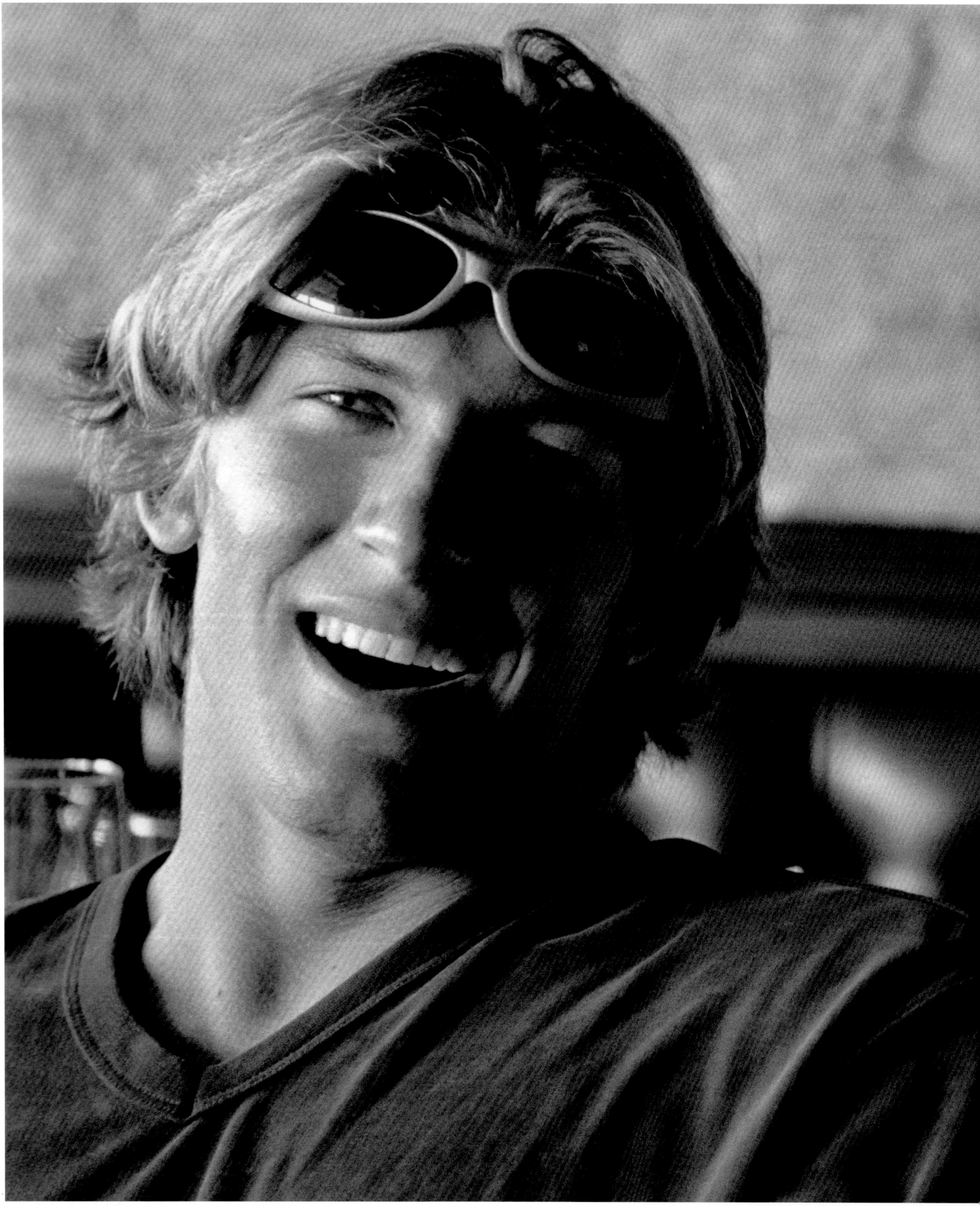

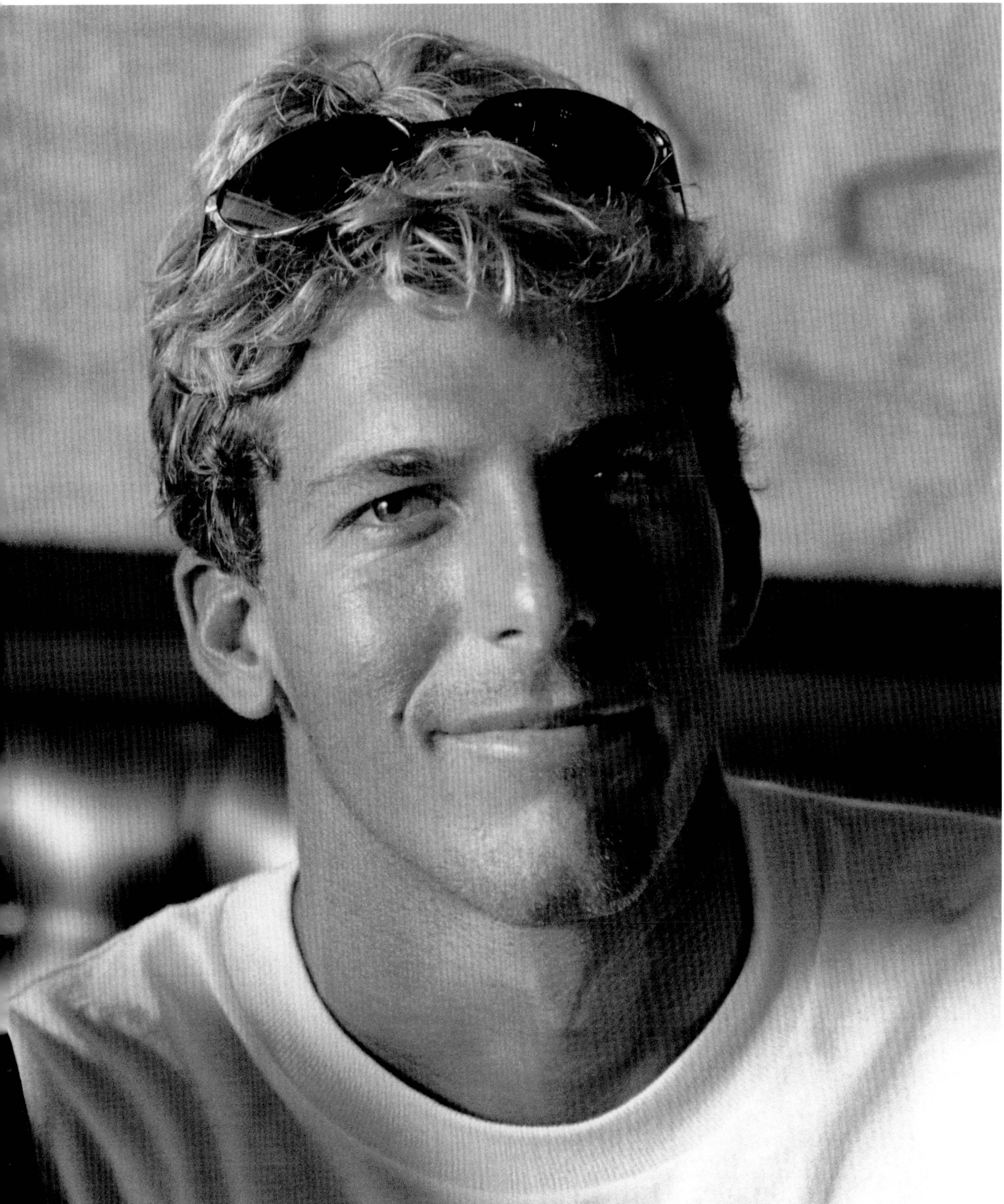

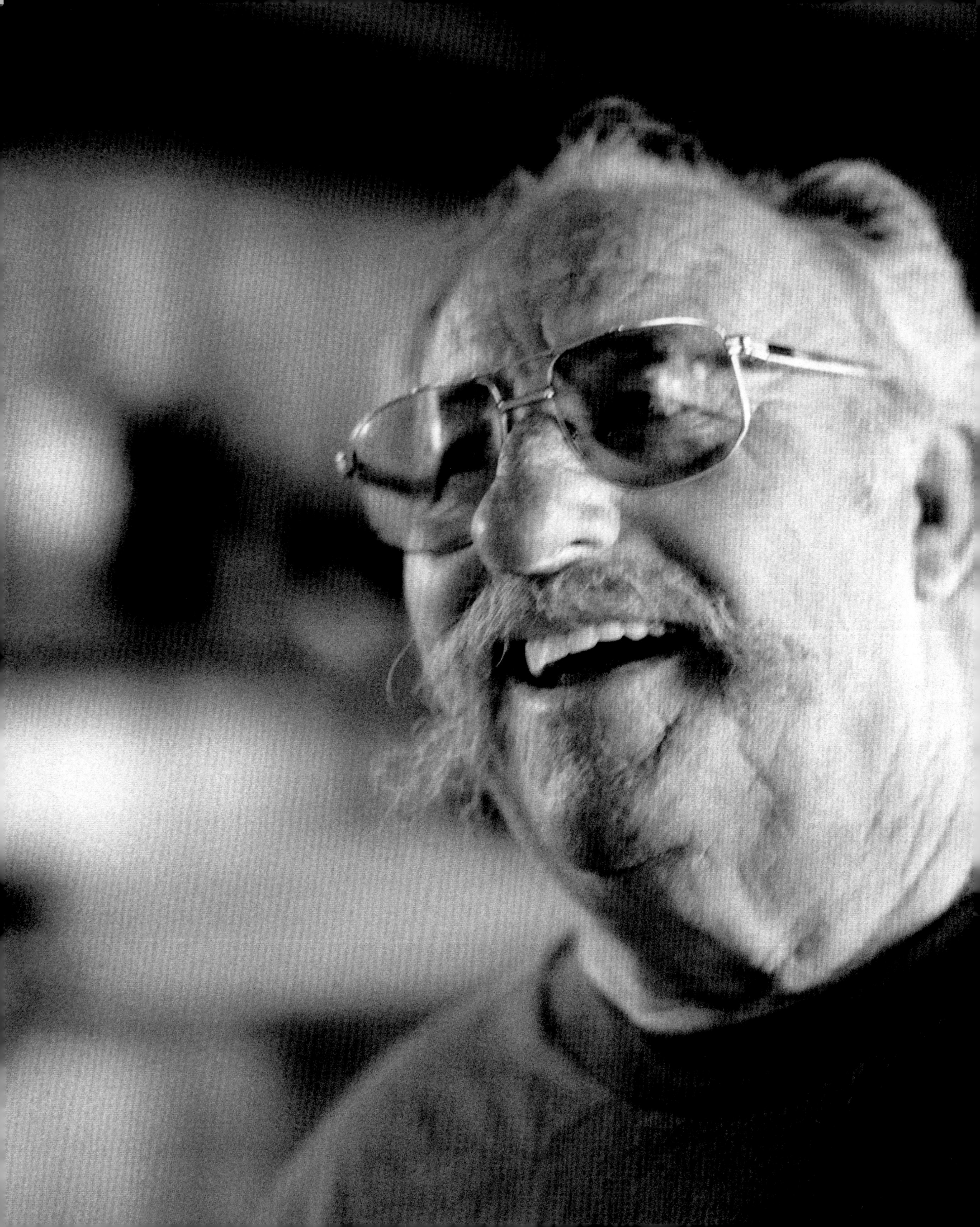

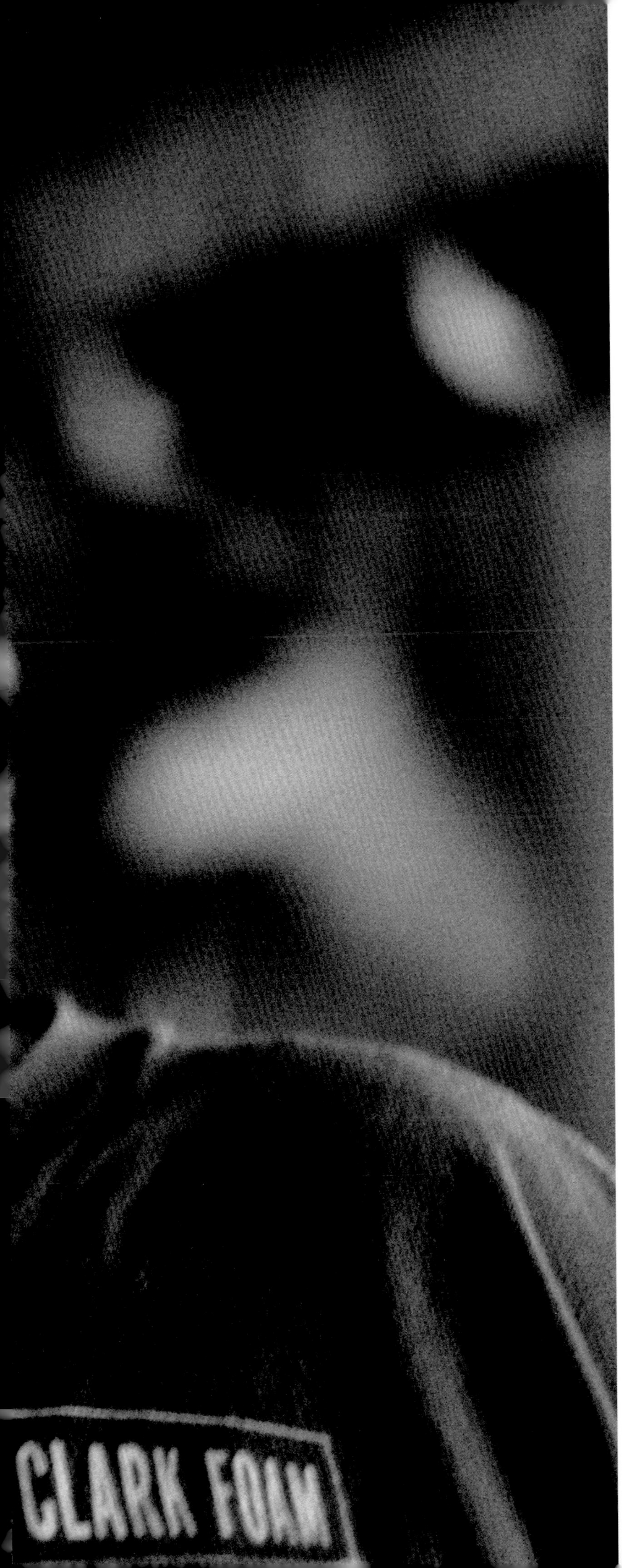
CLARK FOAM

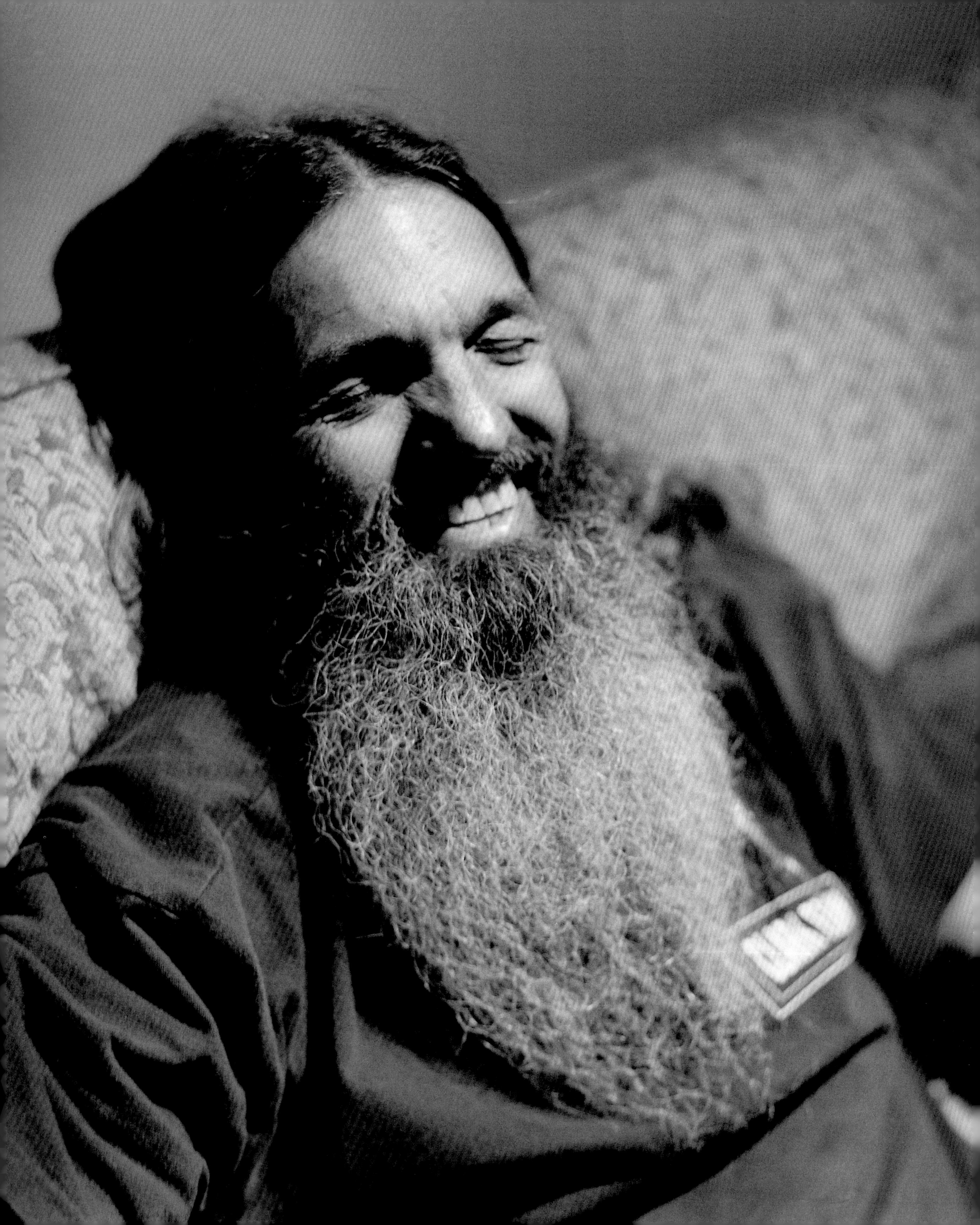

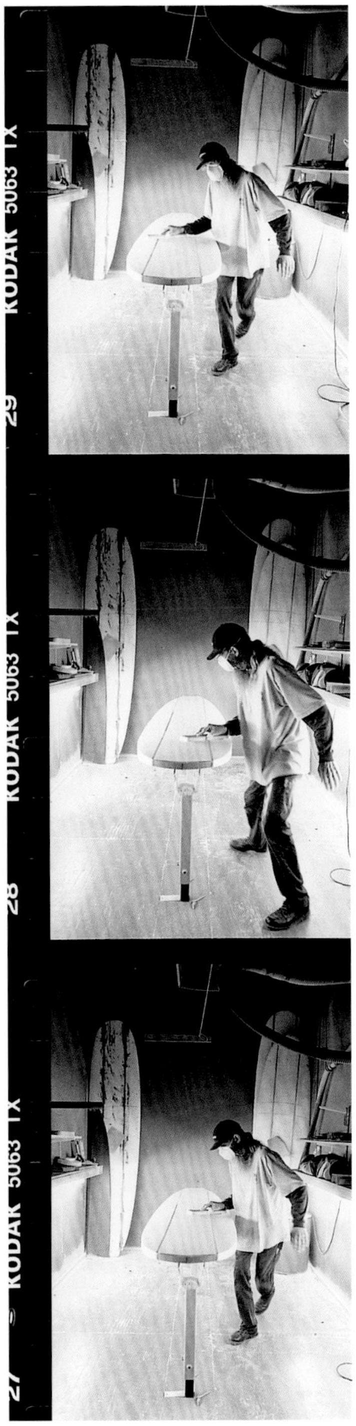
KODAK 5063 TX
29
KODAK 5063 TX
28
KODAK 5063 TX
27

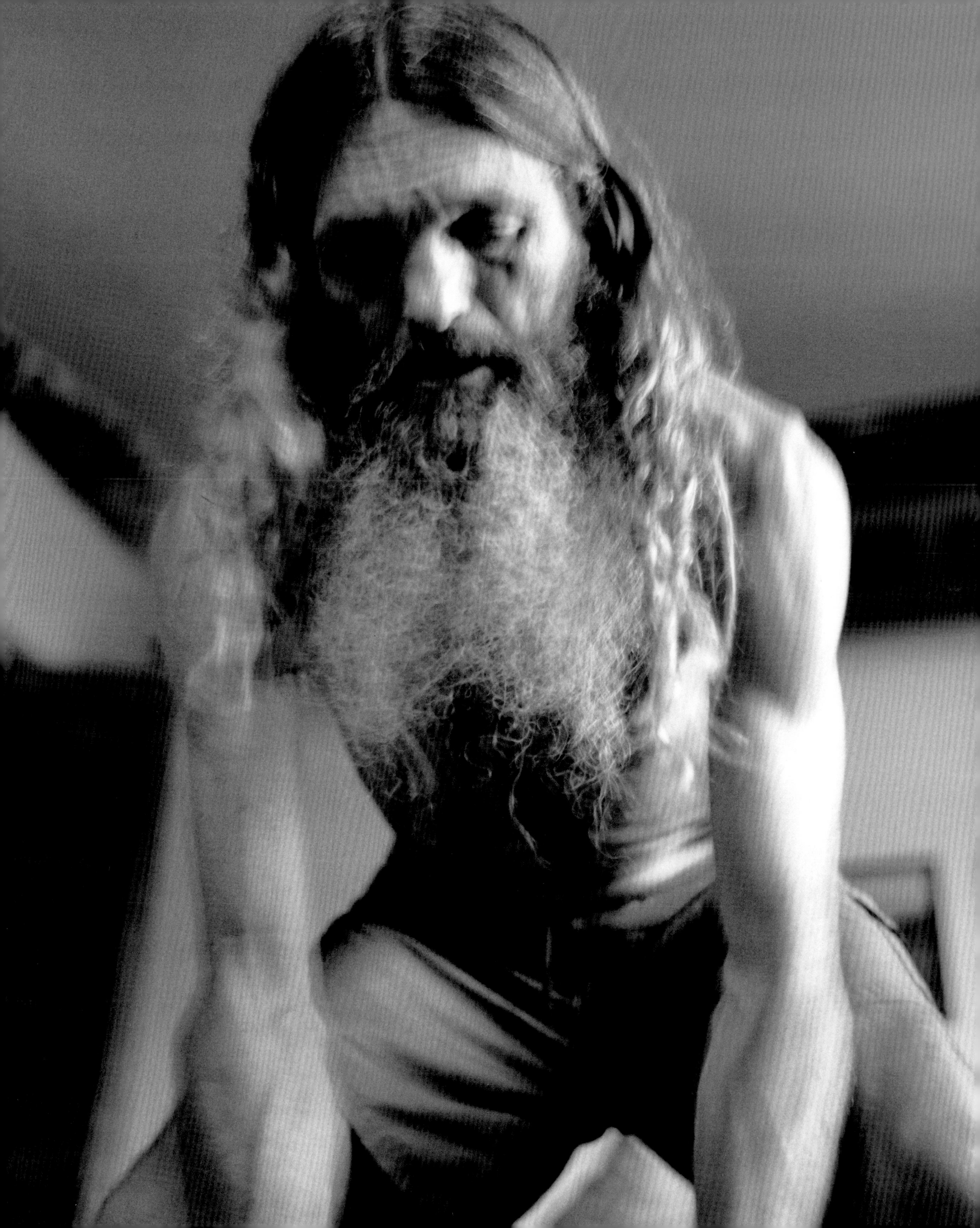

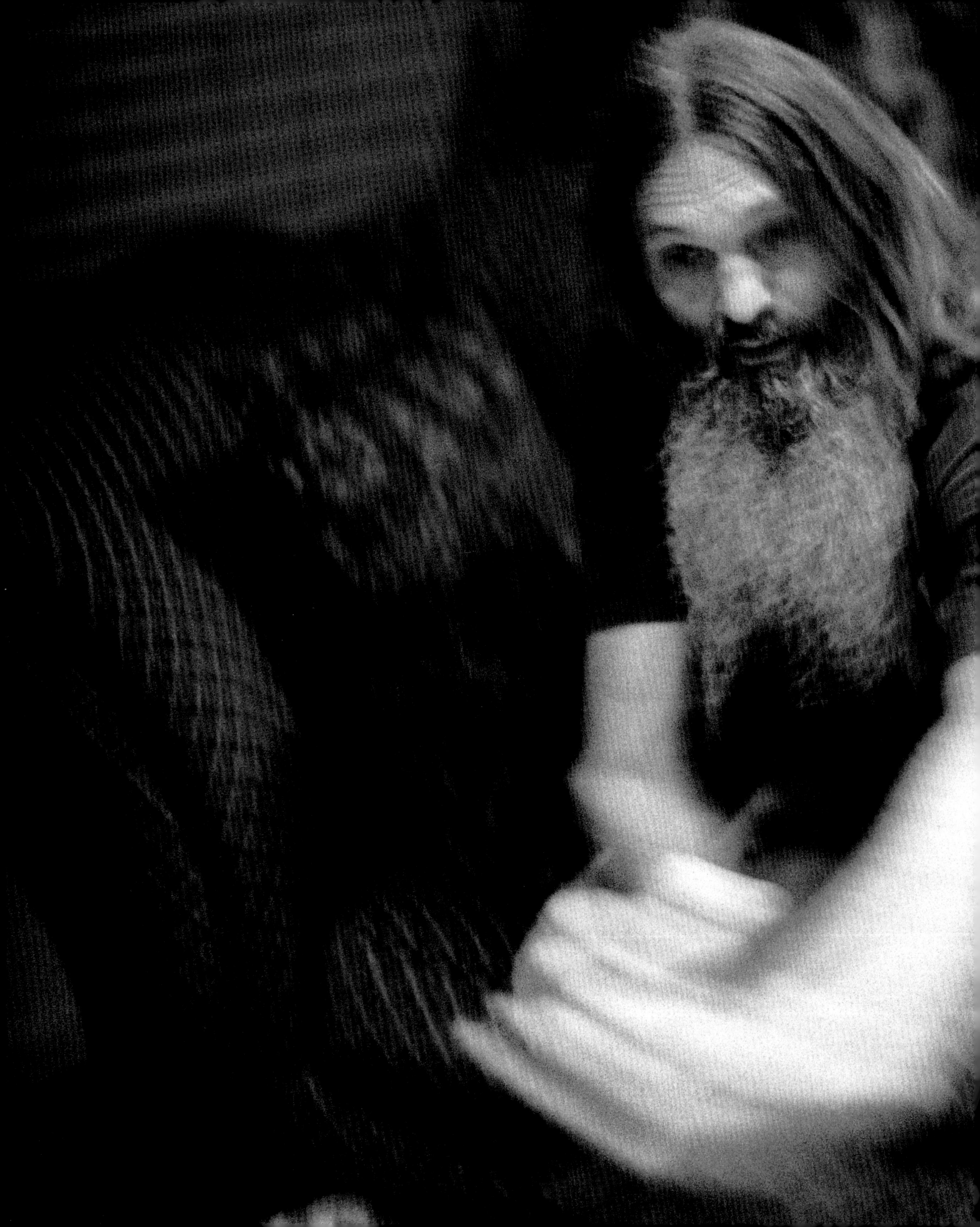

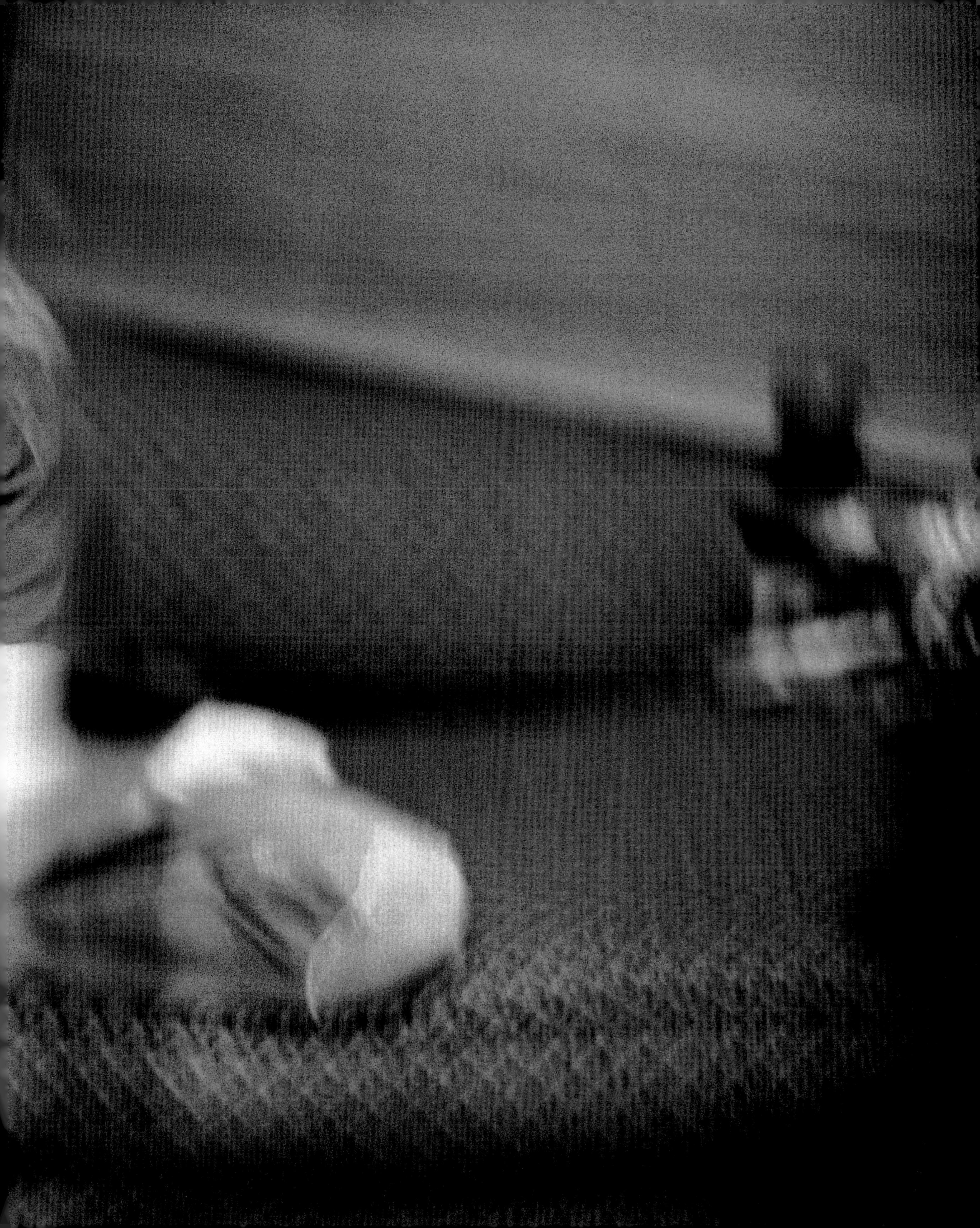

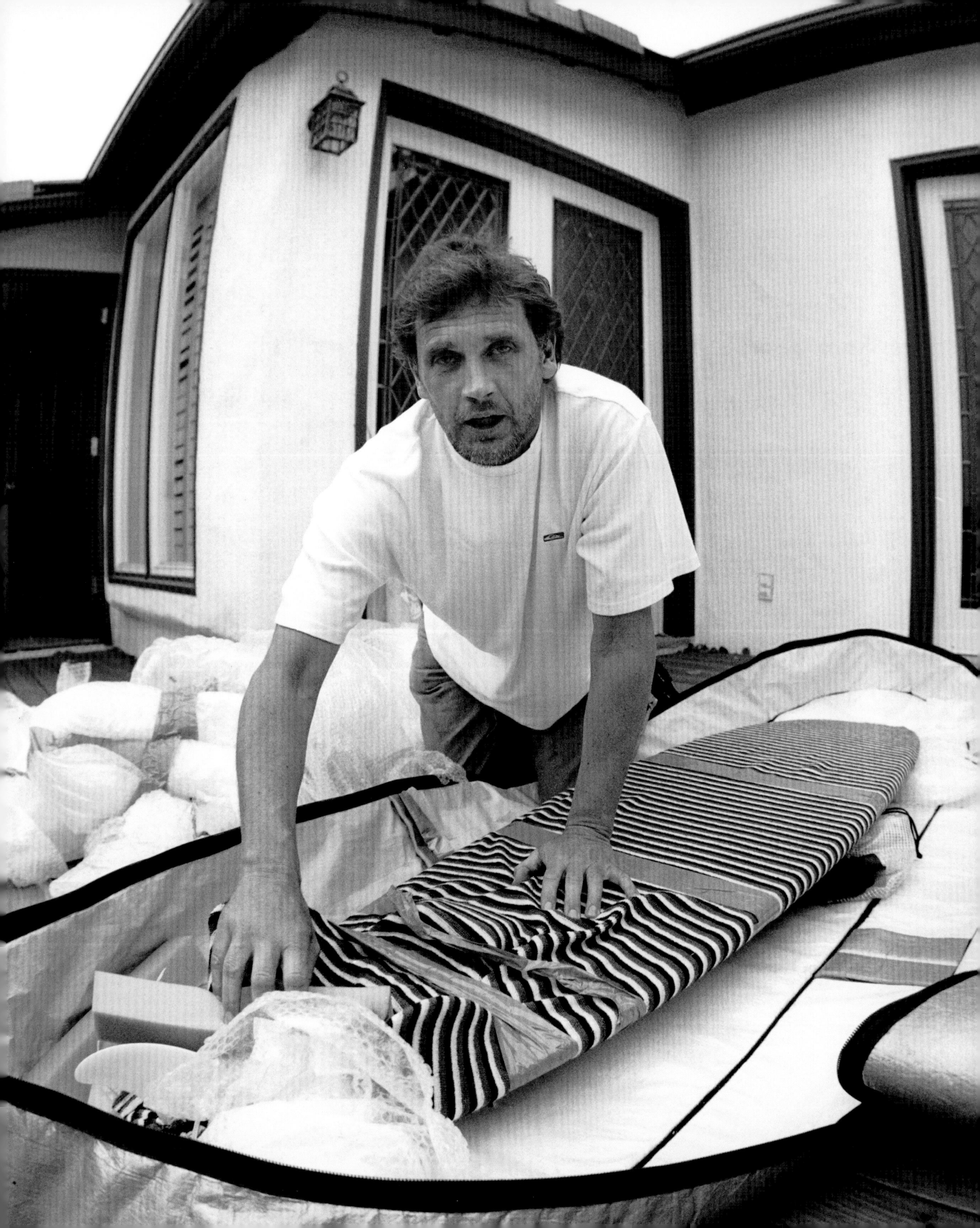

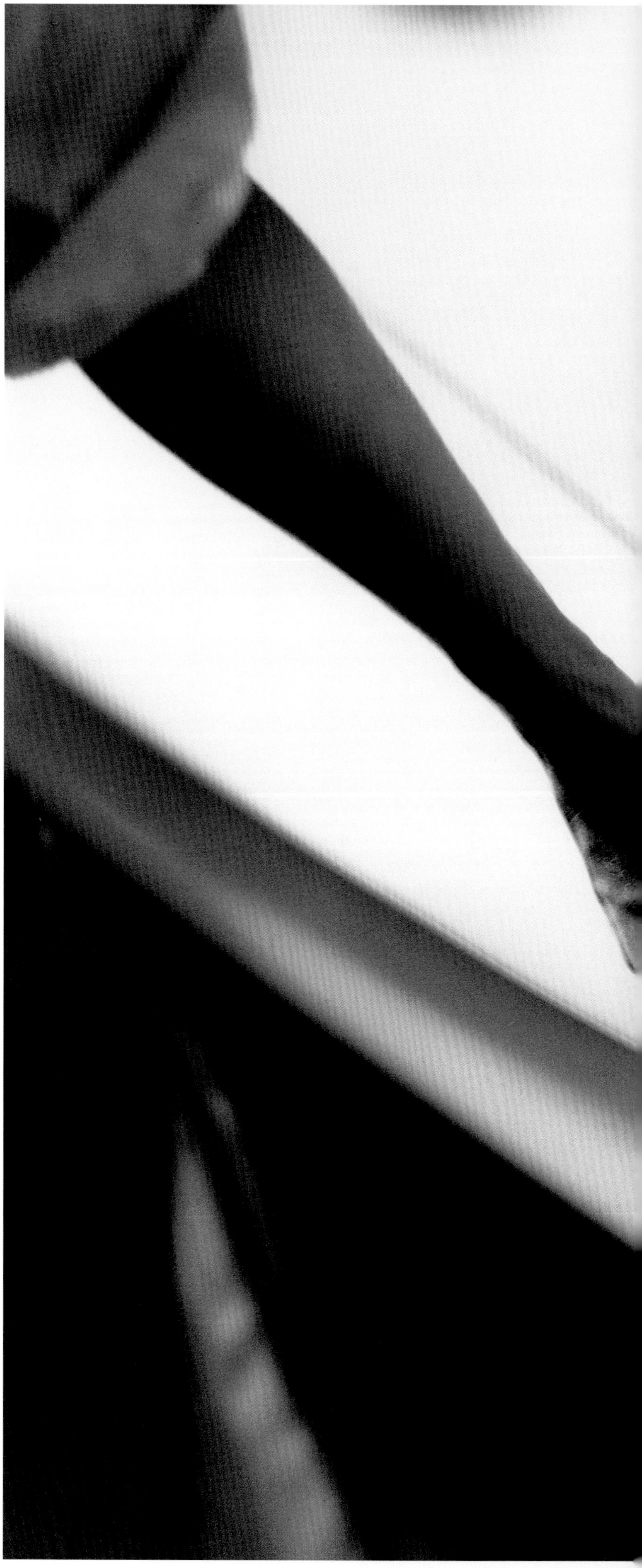

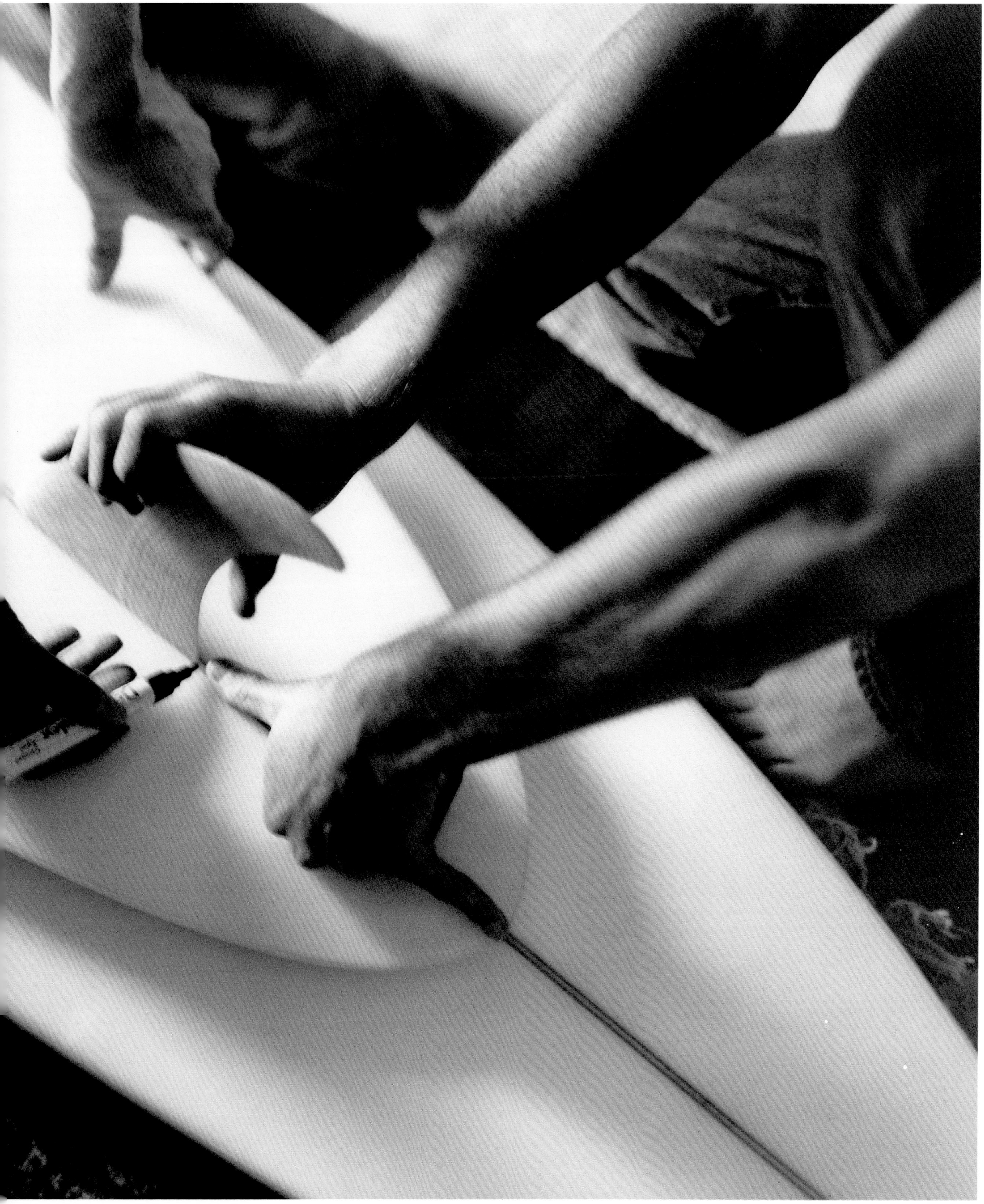

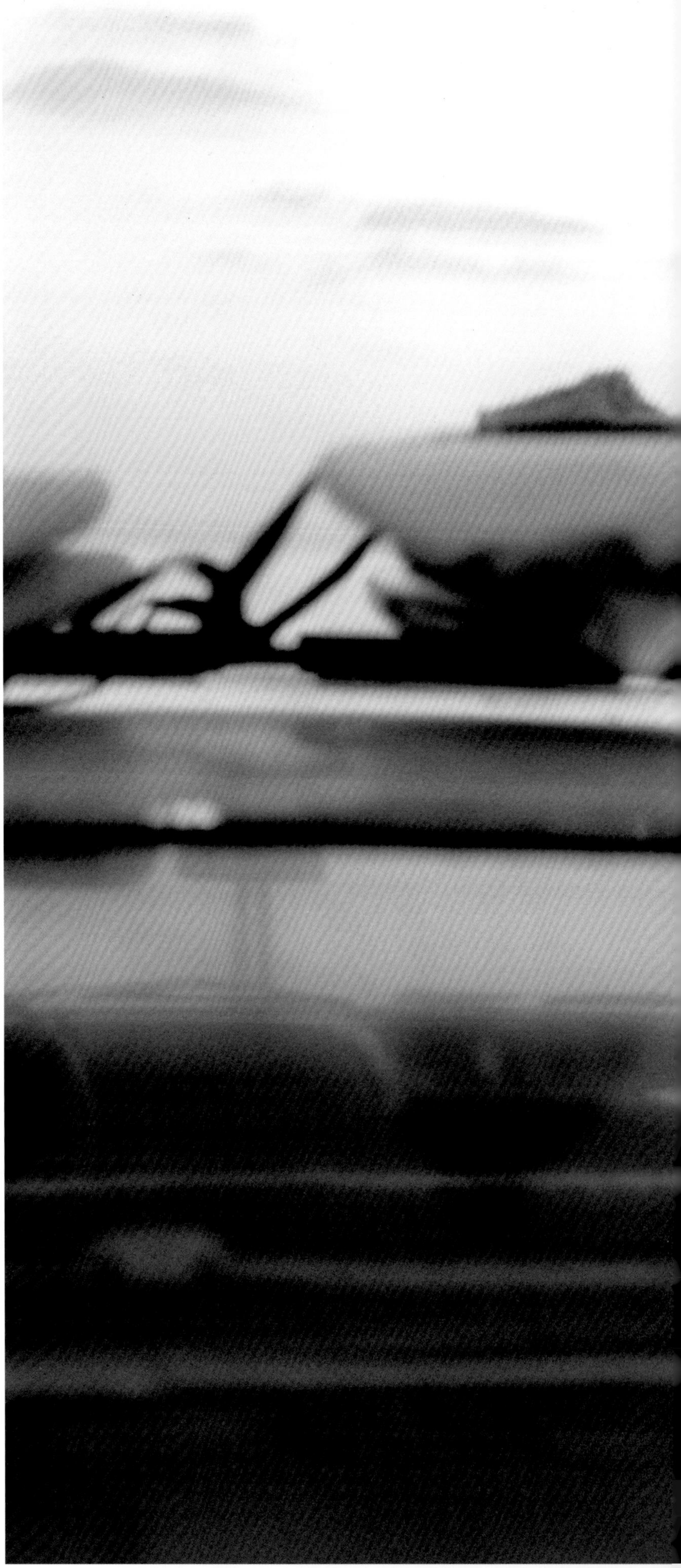

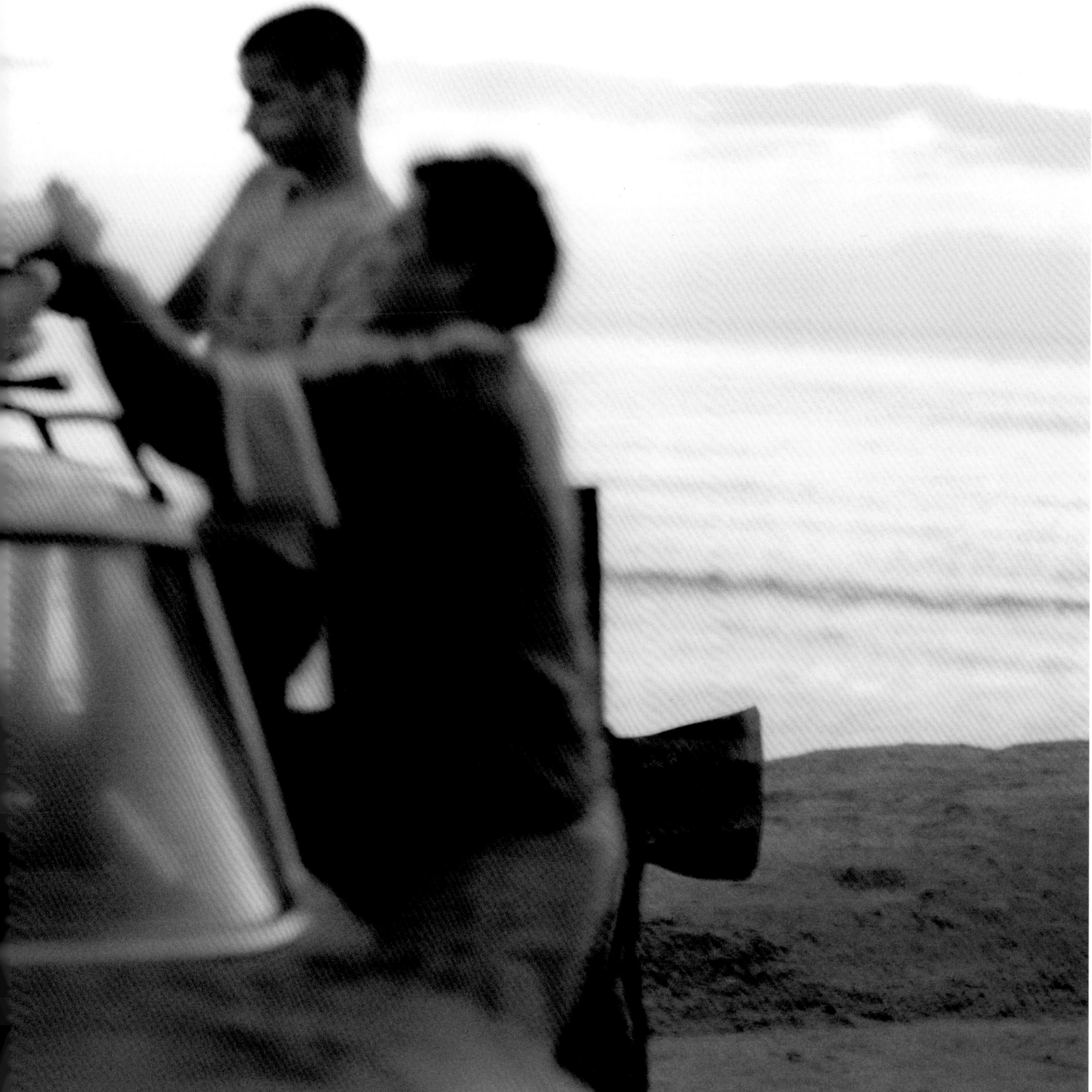

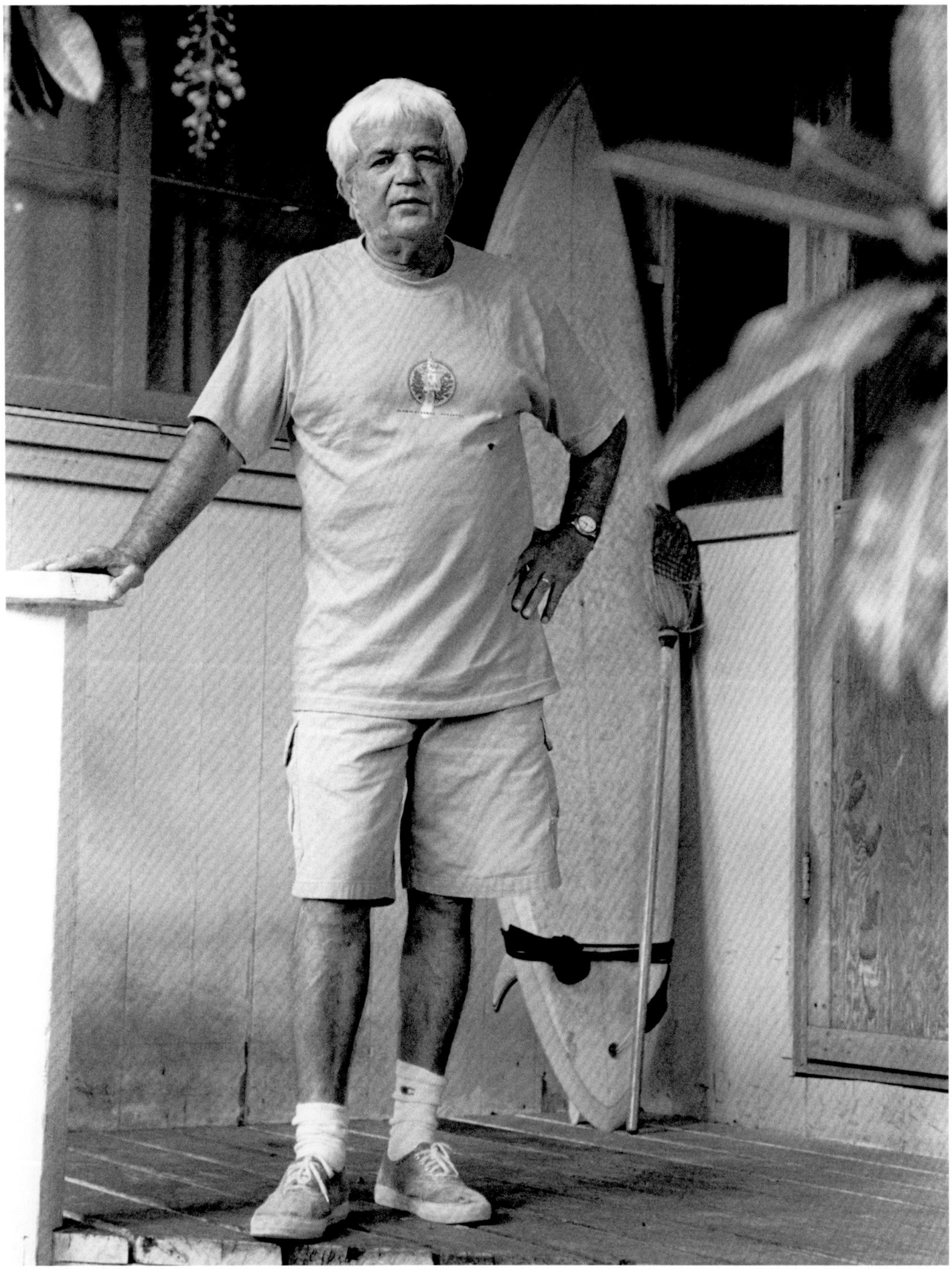

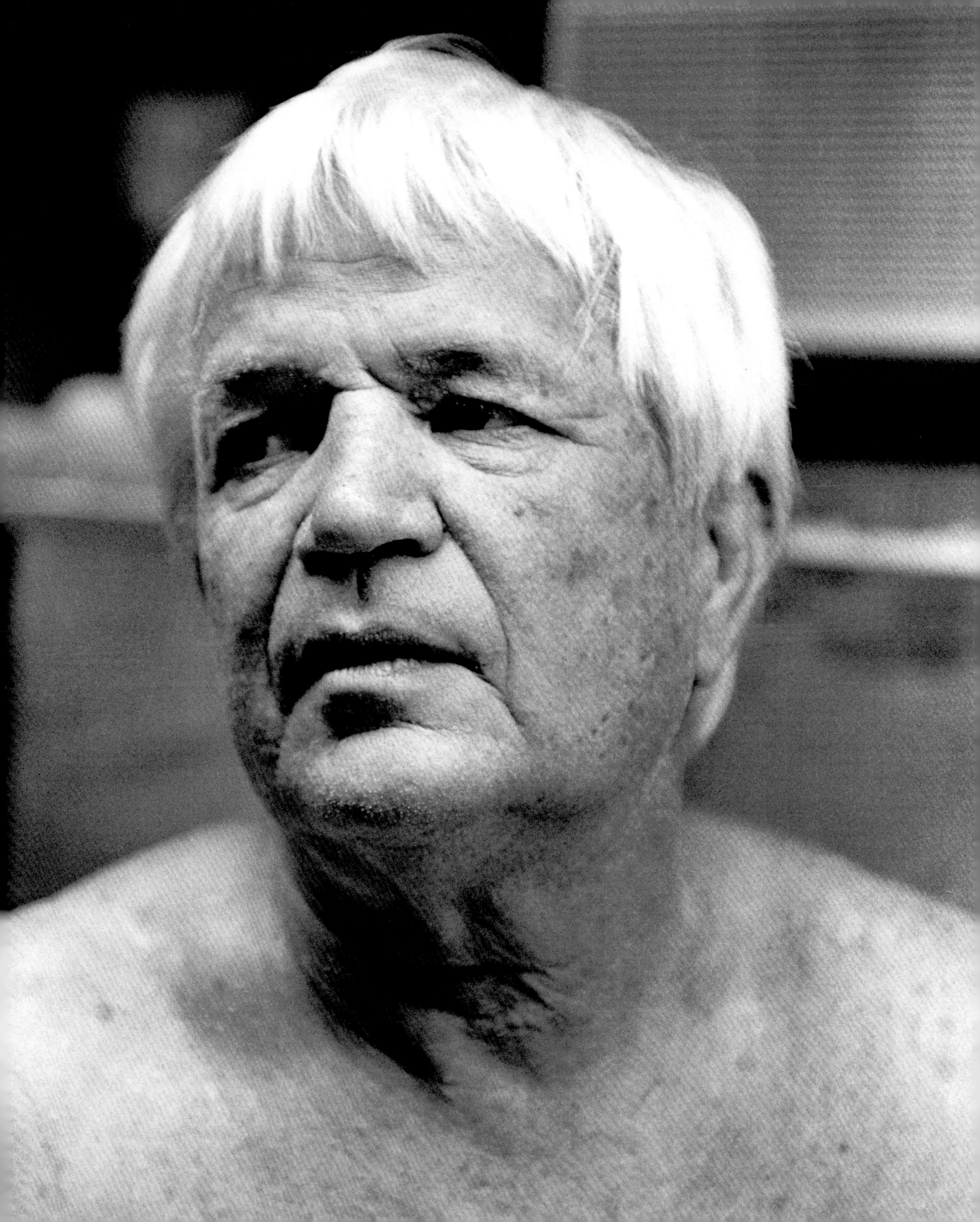

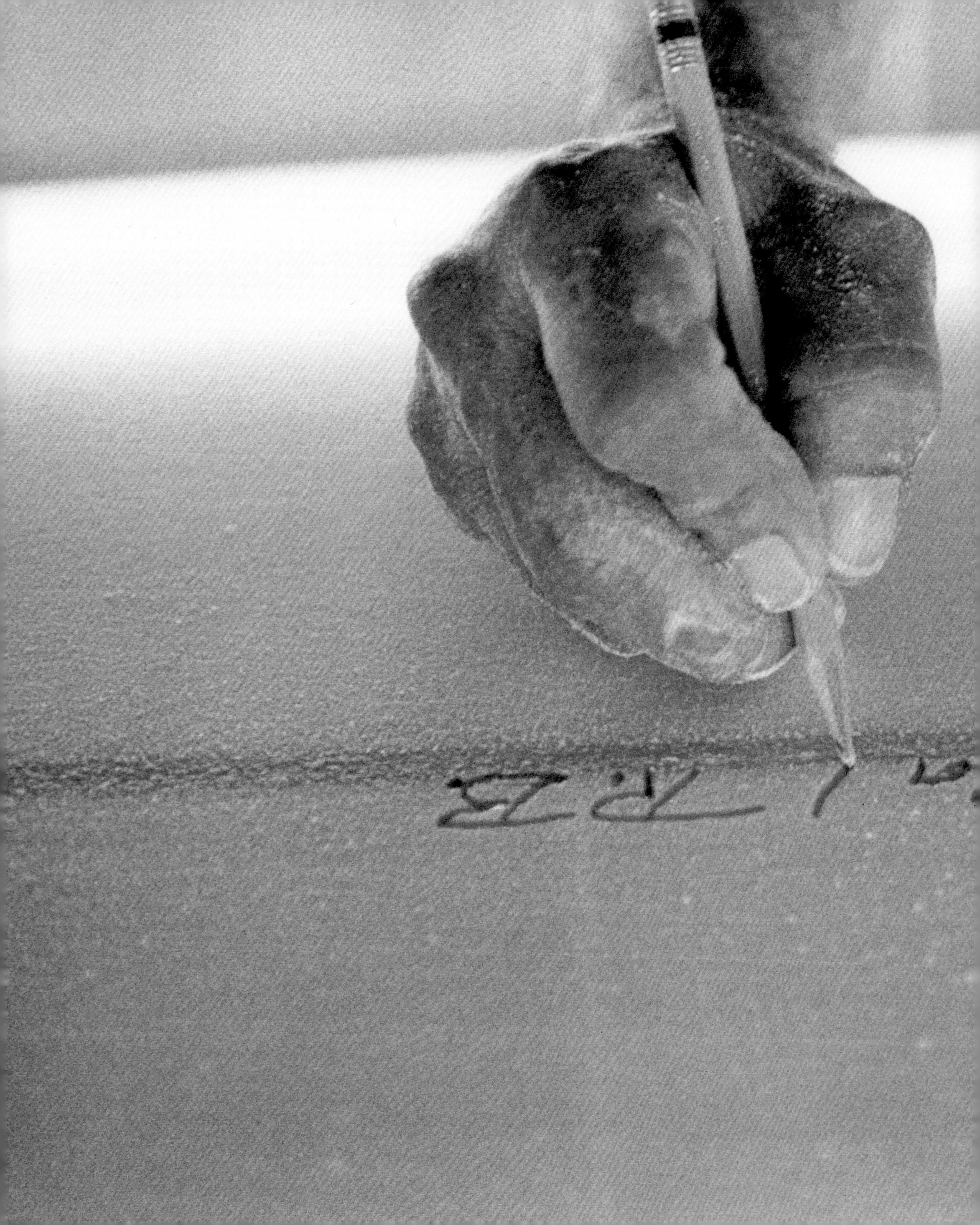

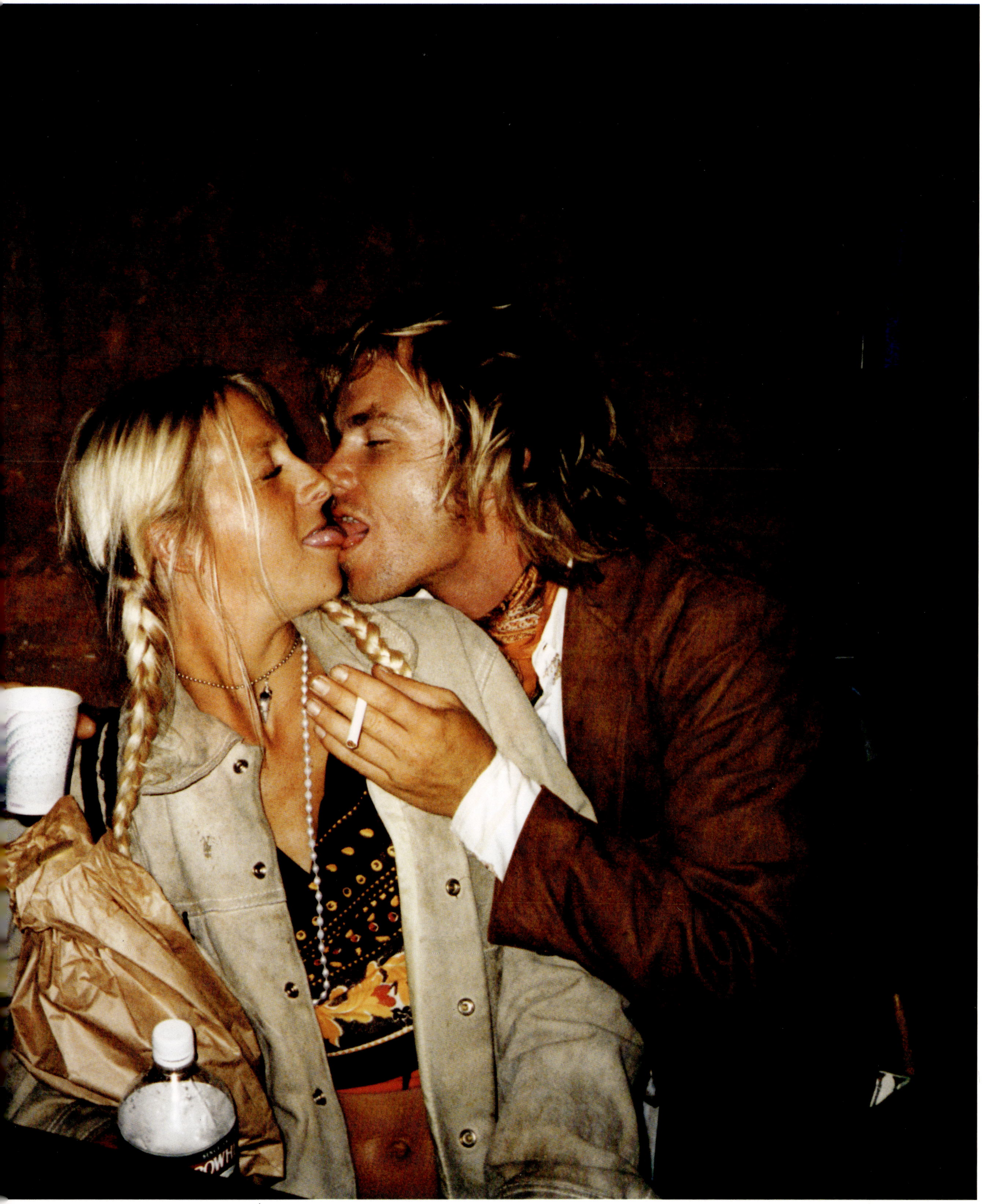

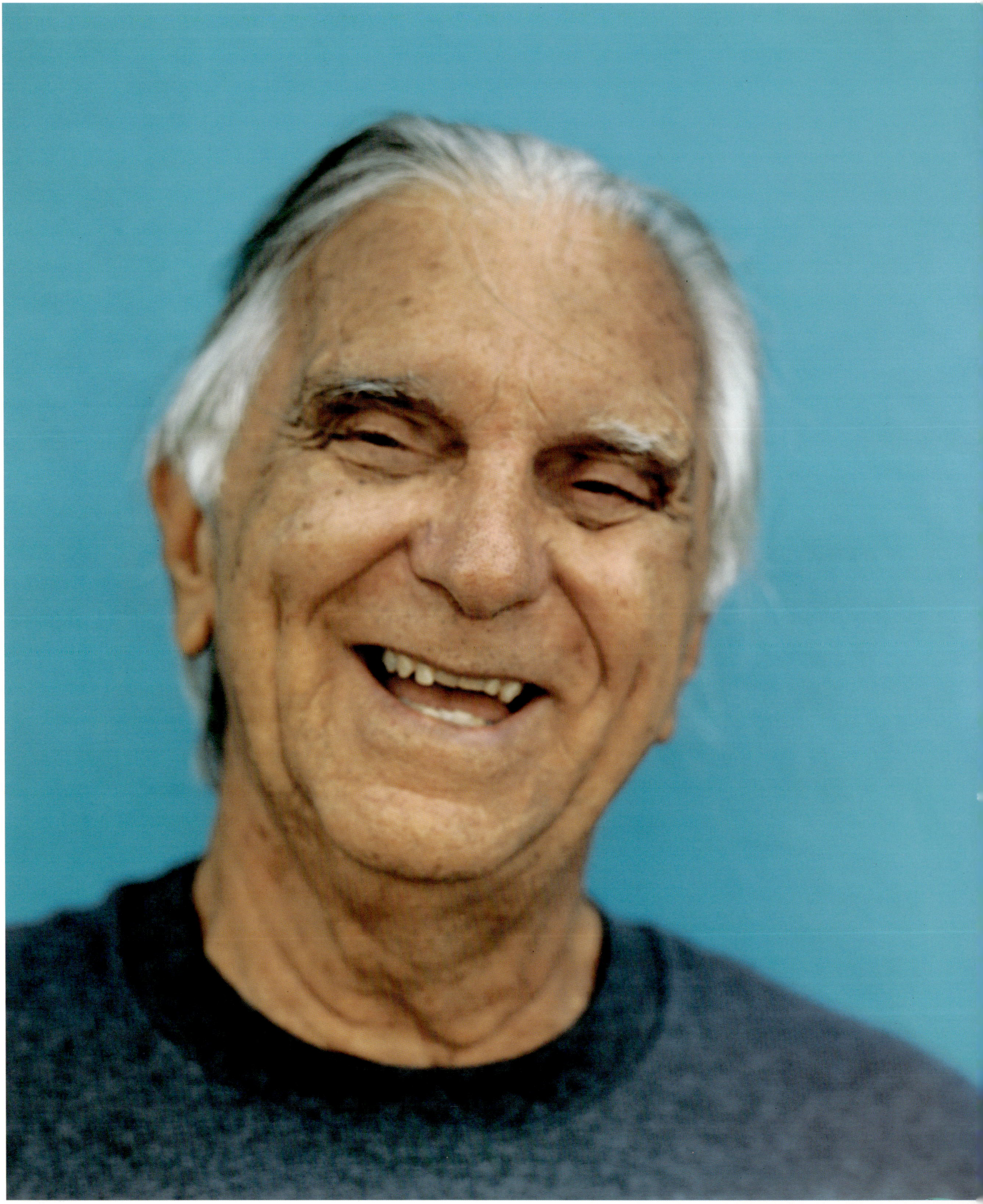

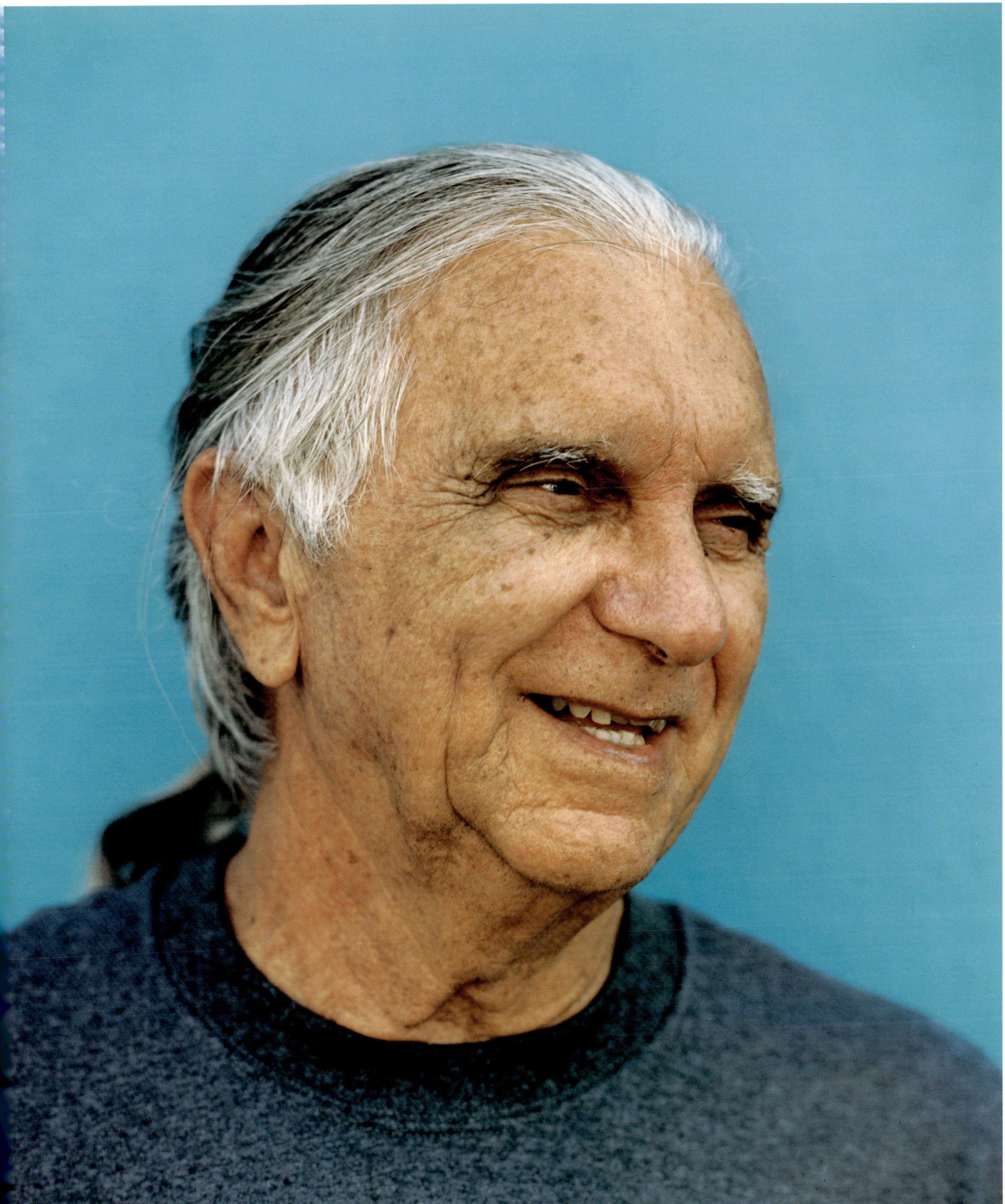

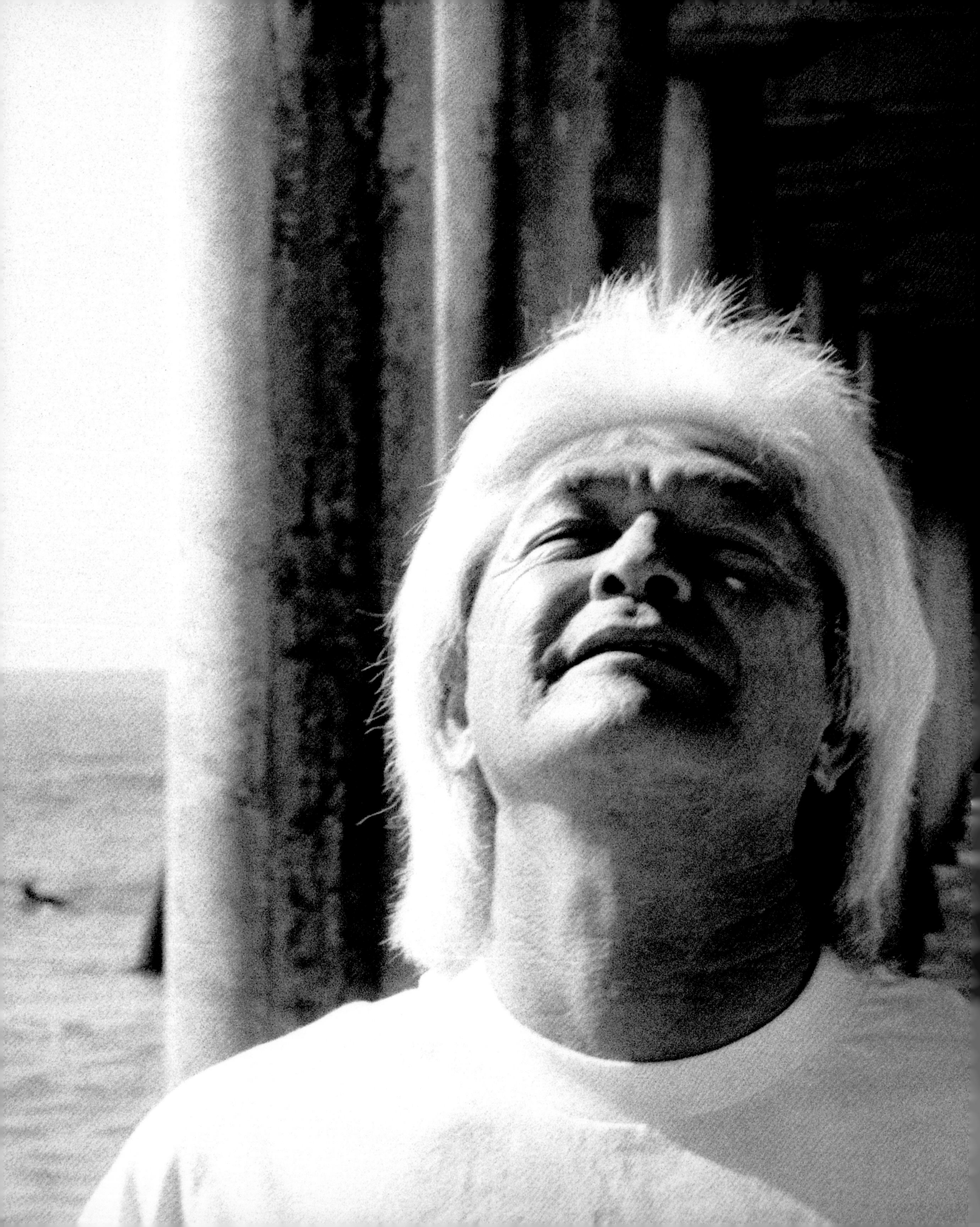

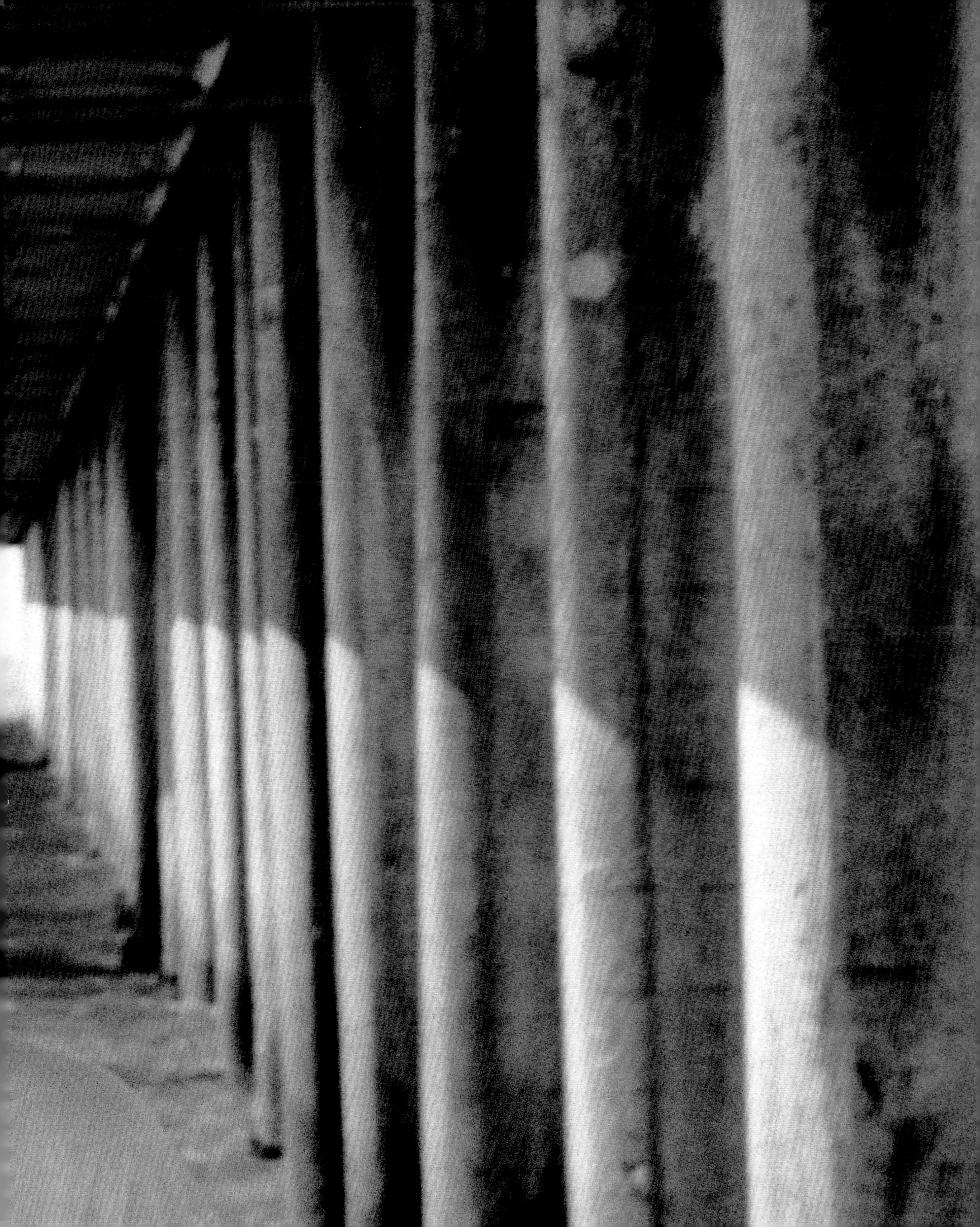

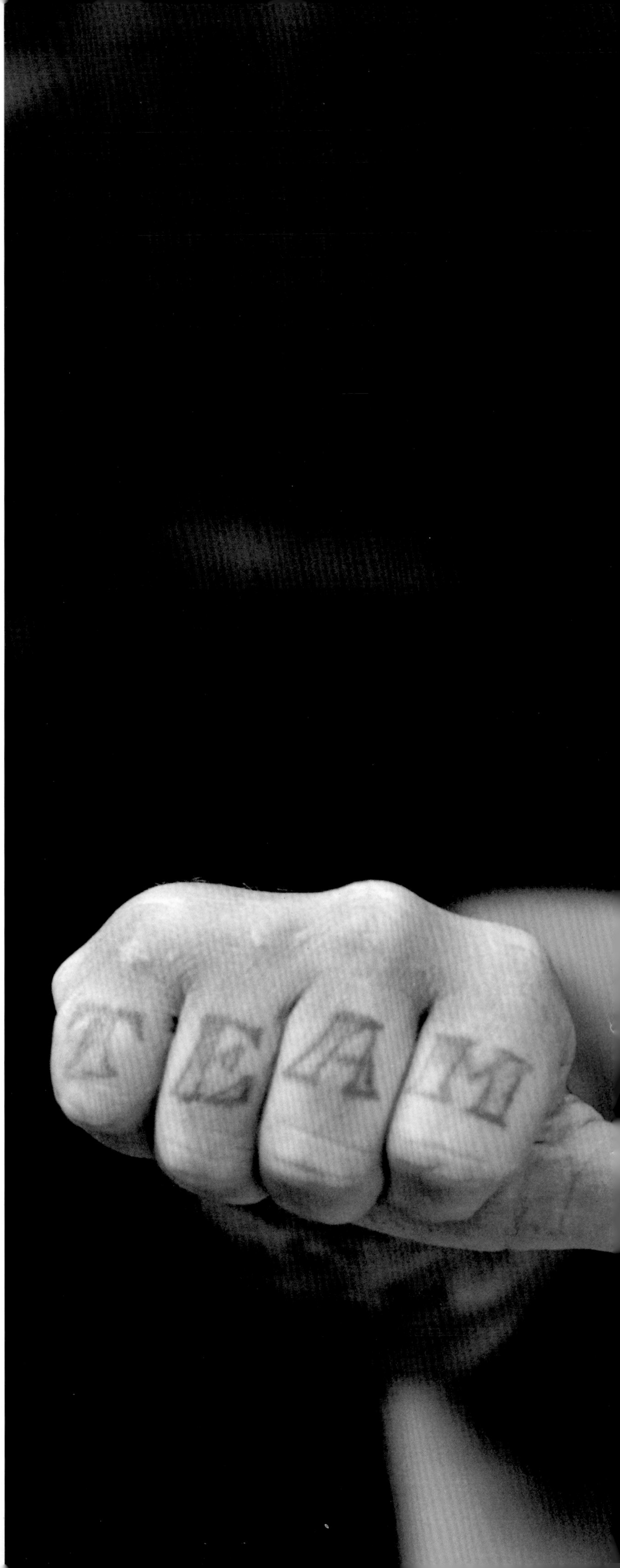
TEAM

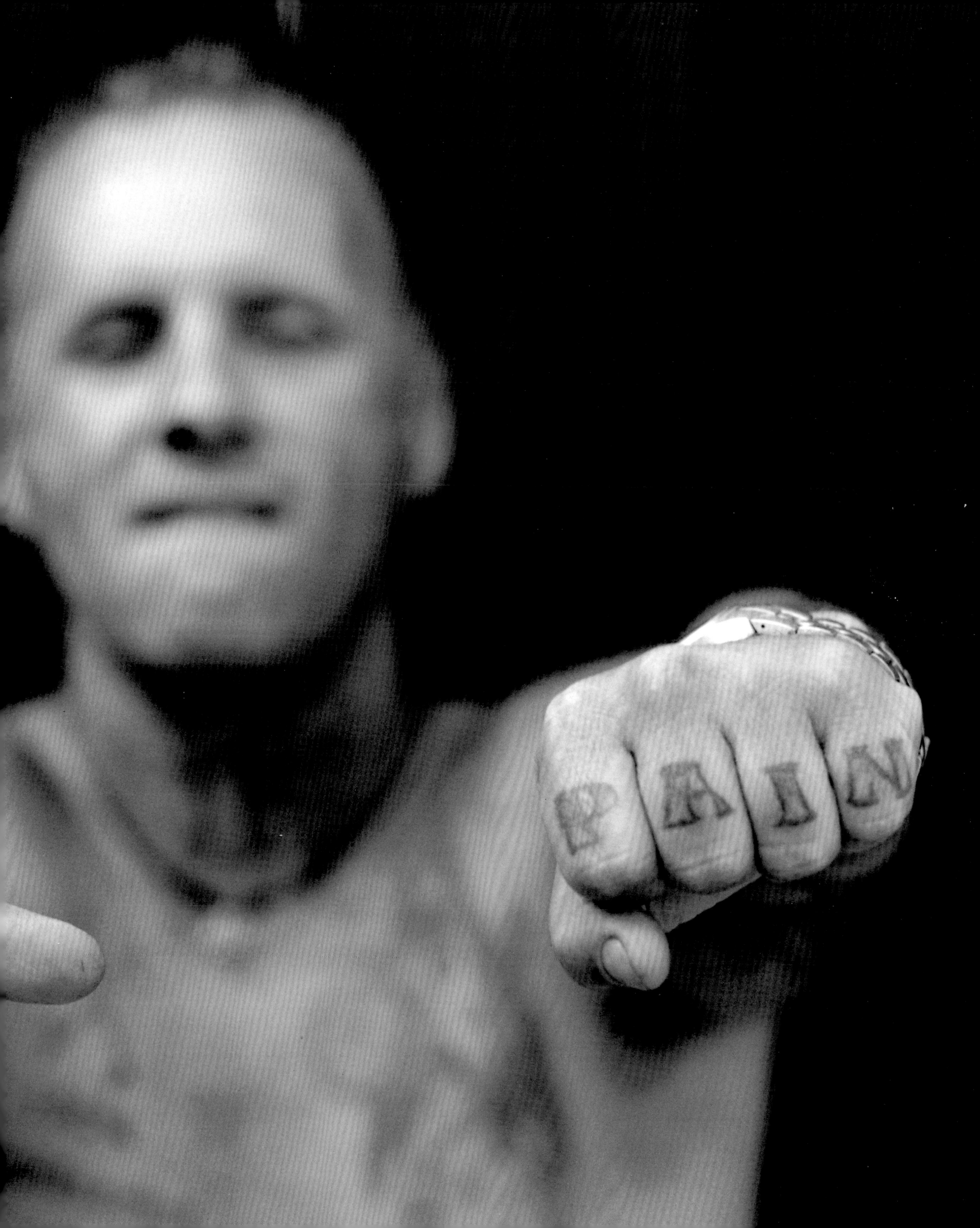
PAIN

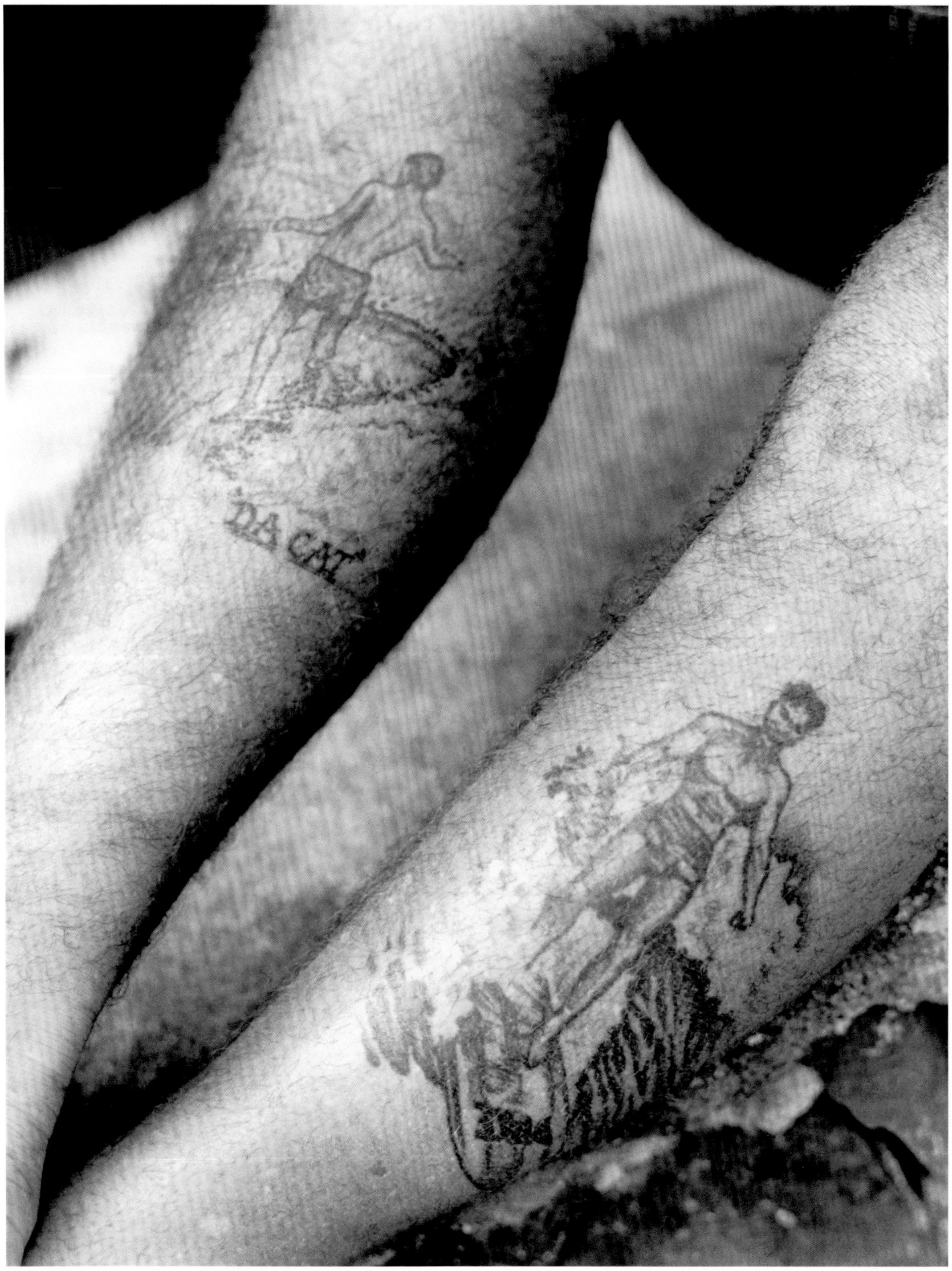
DA CAT

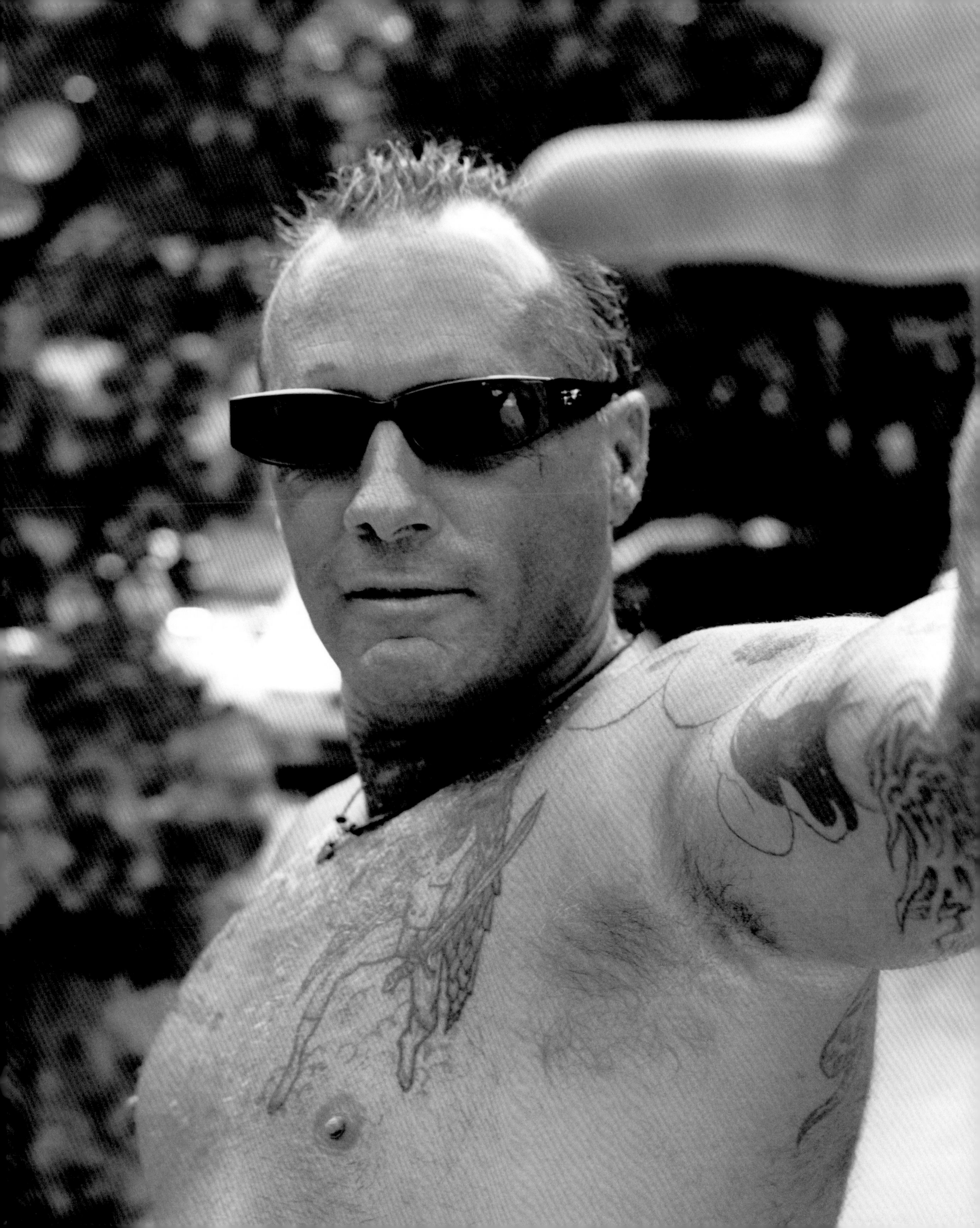

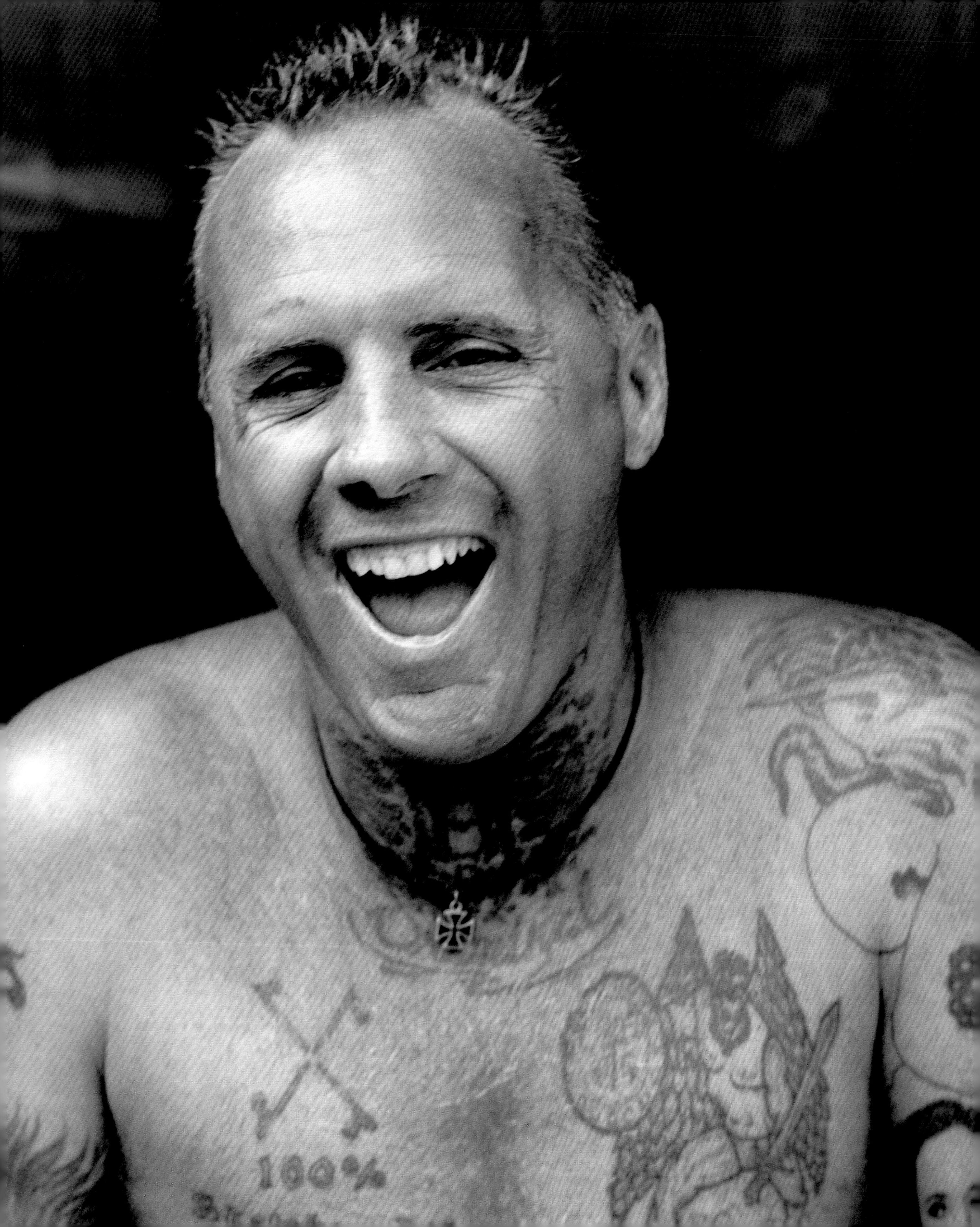
100%

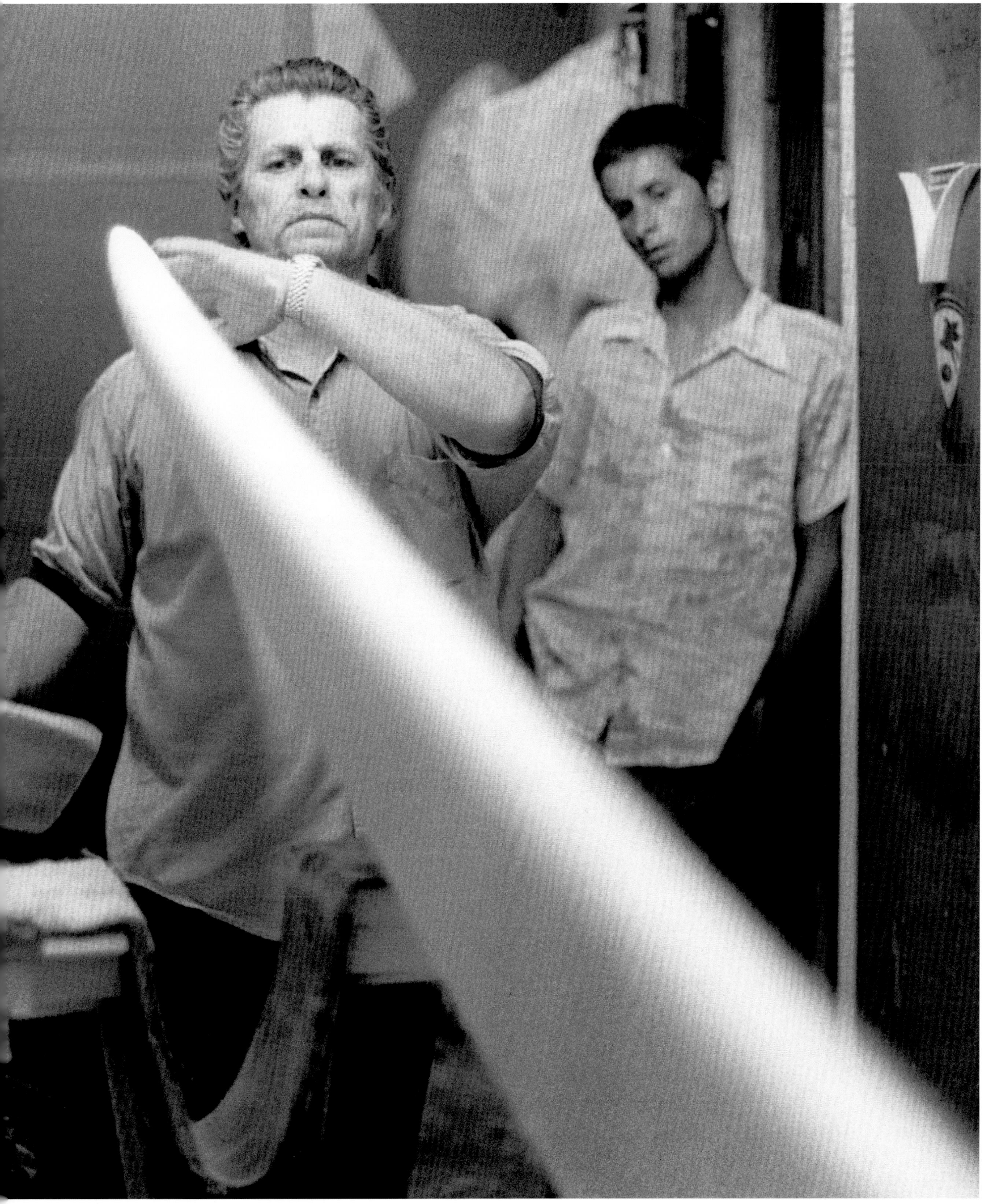

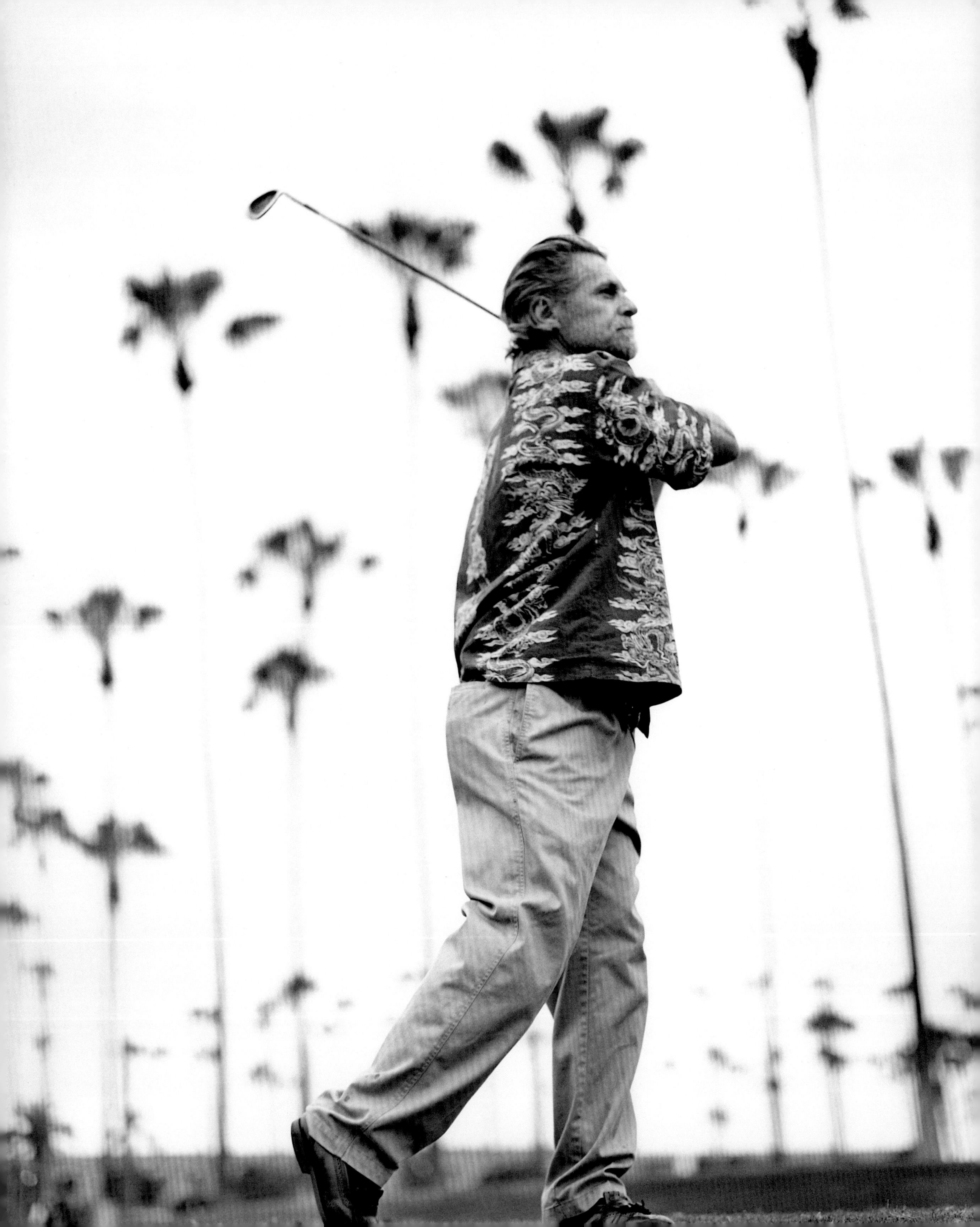

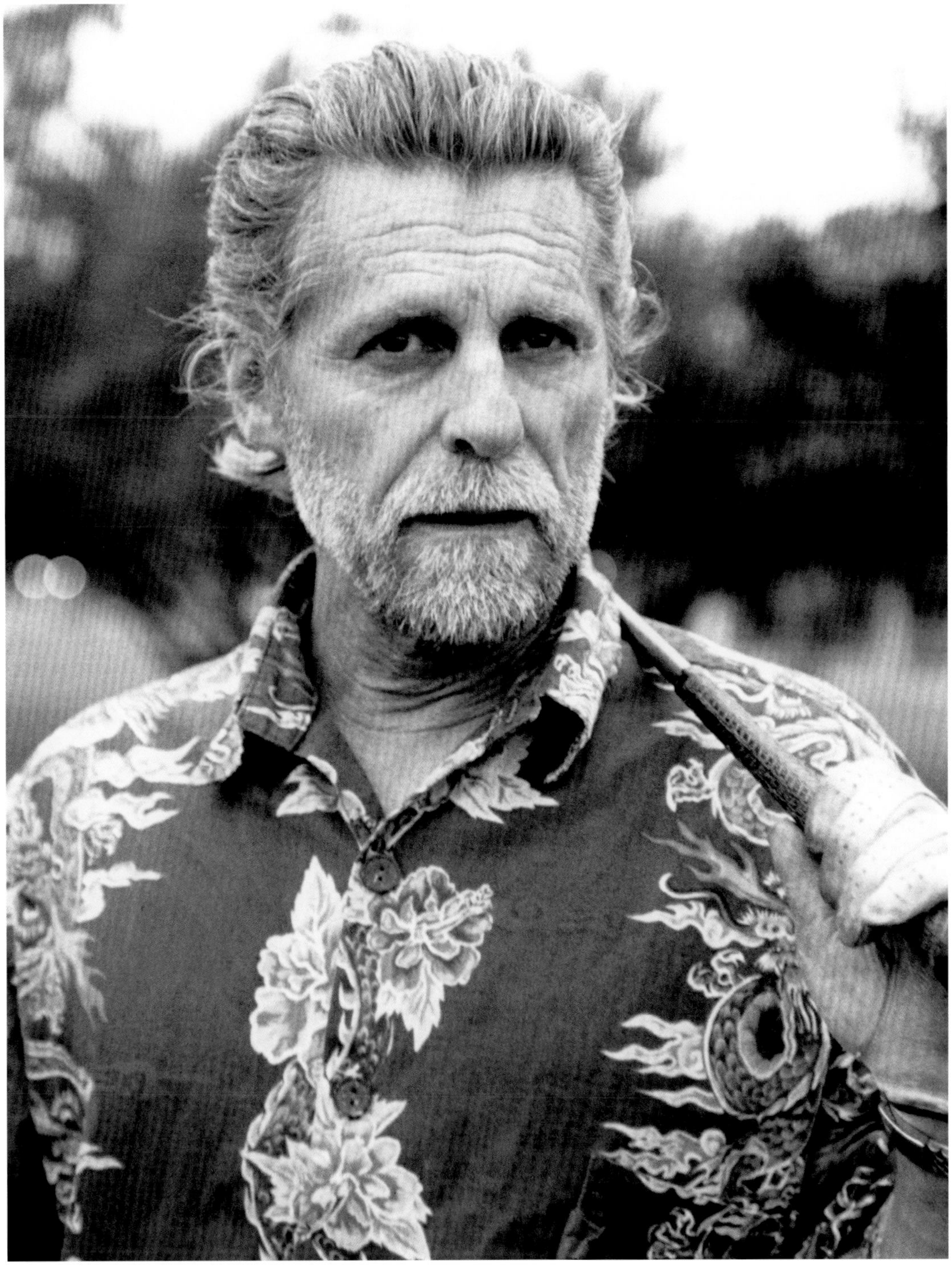

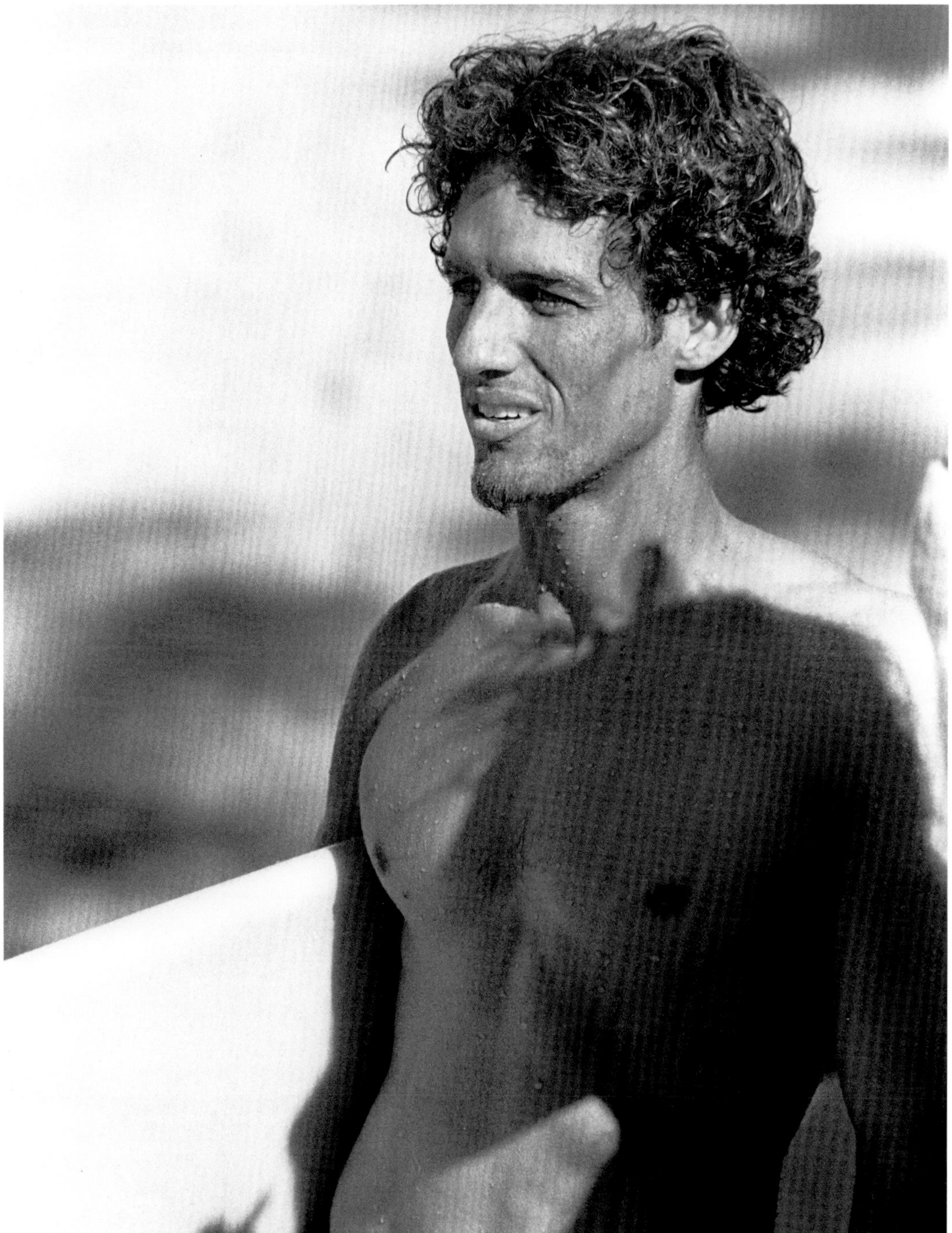

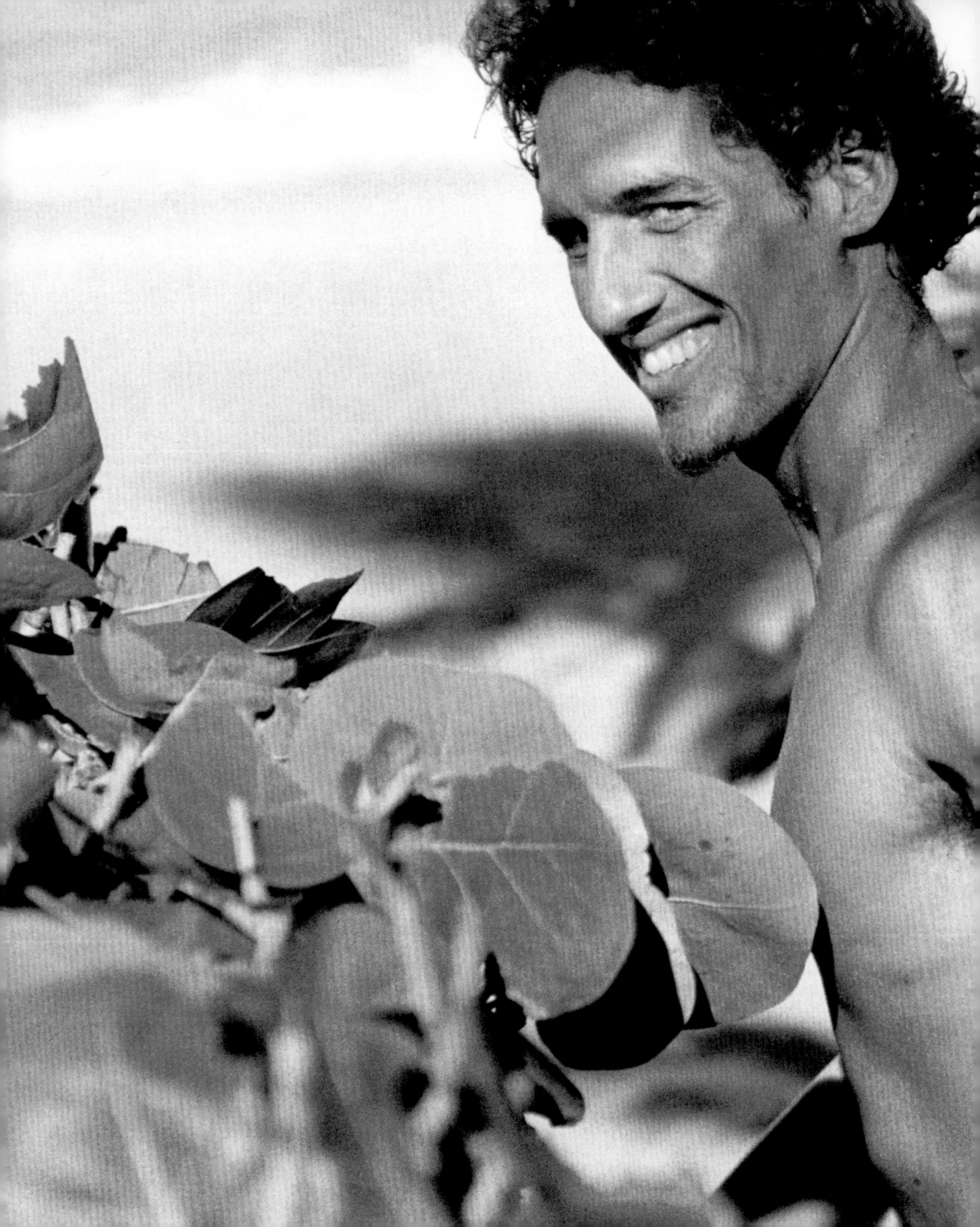

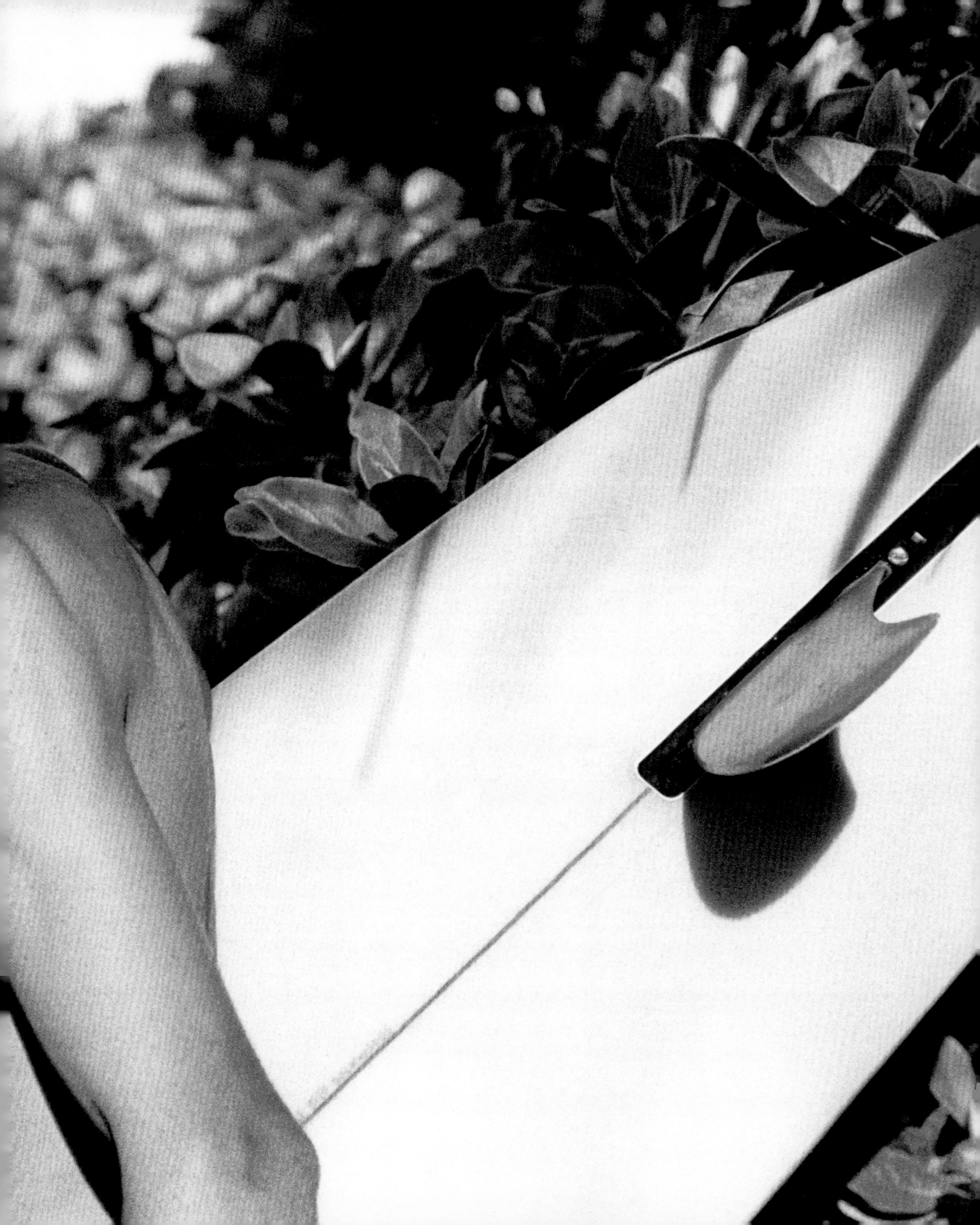

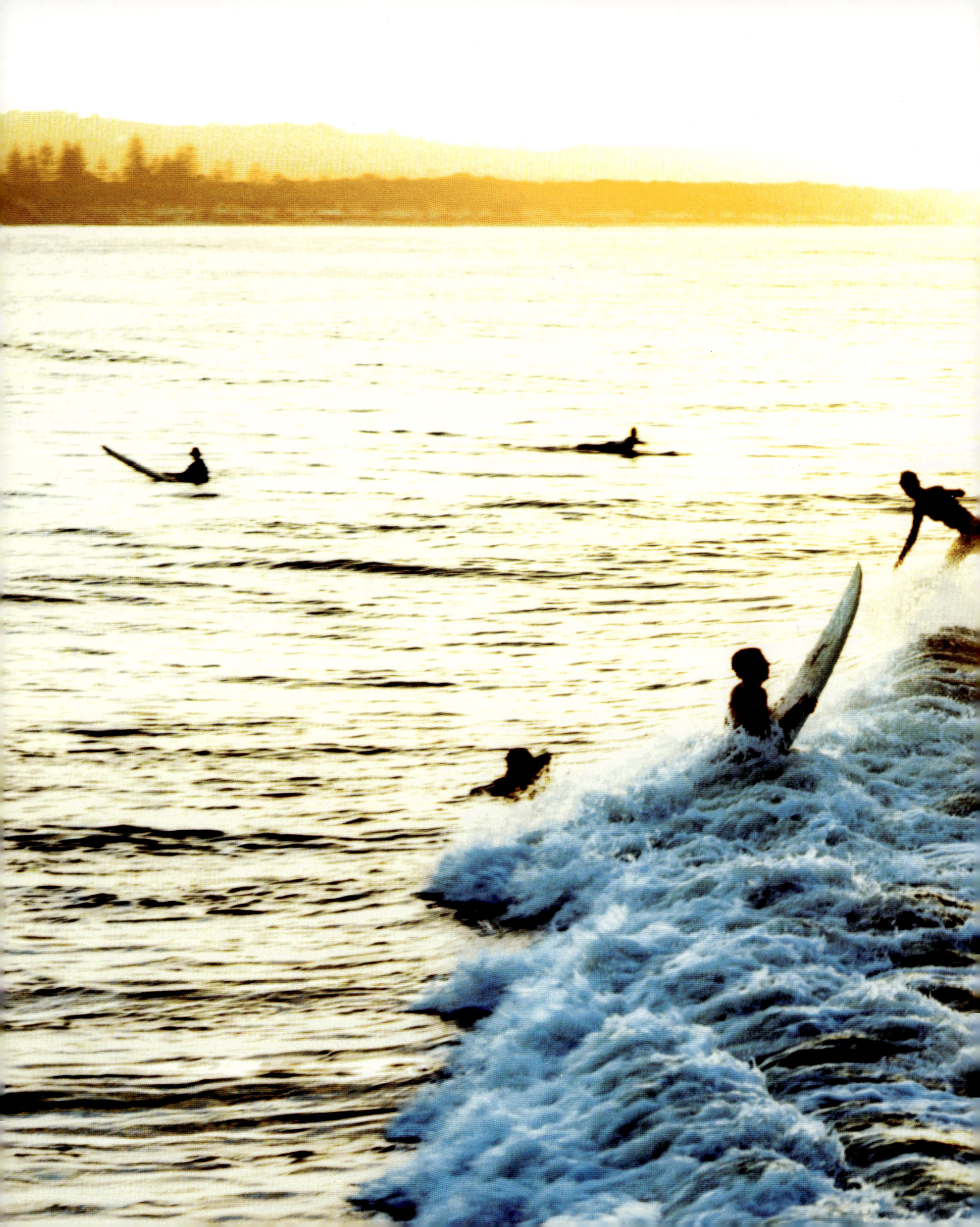

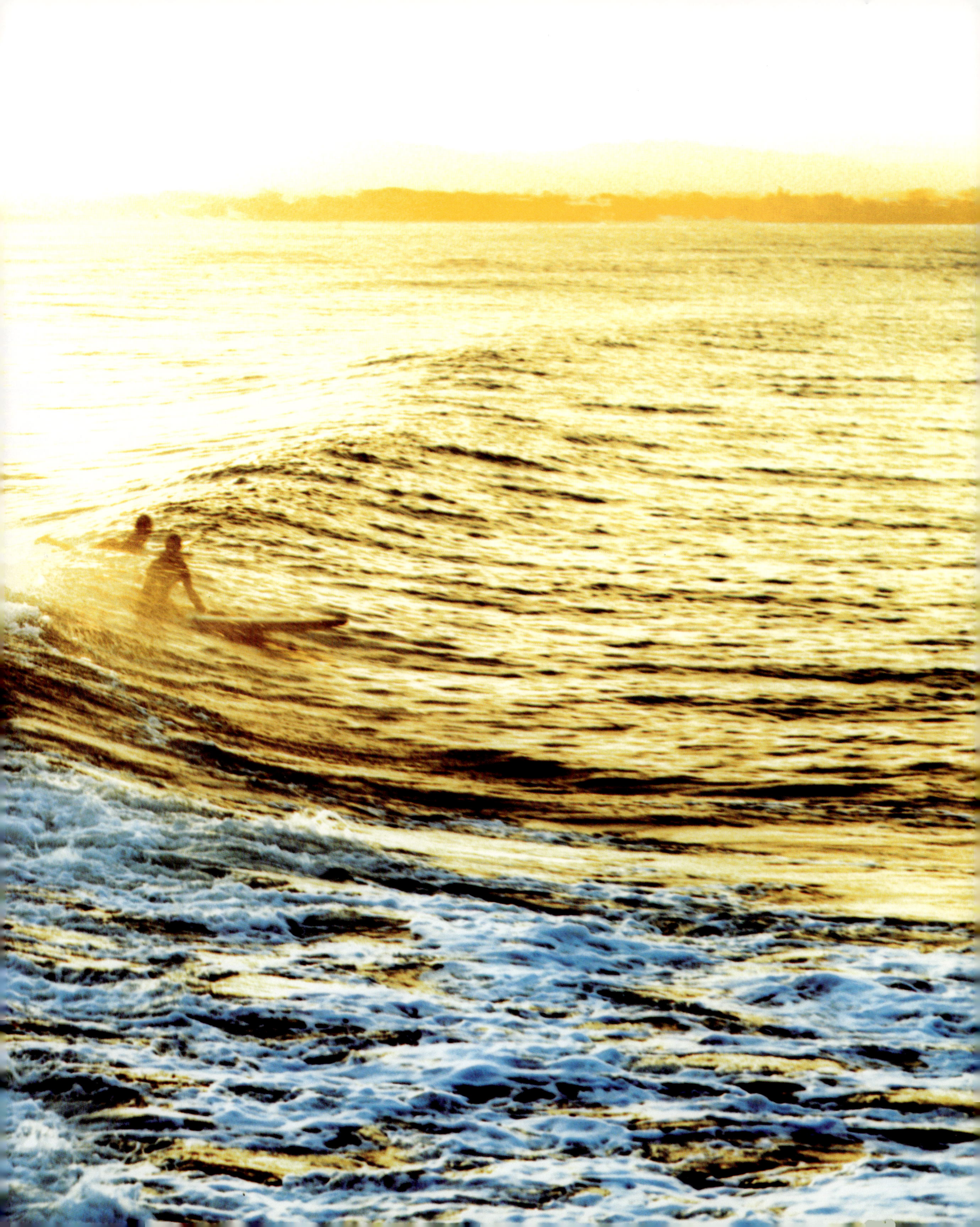

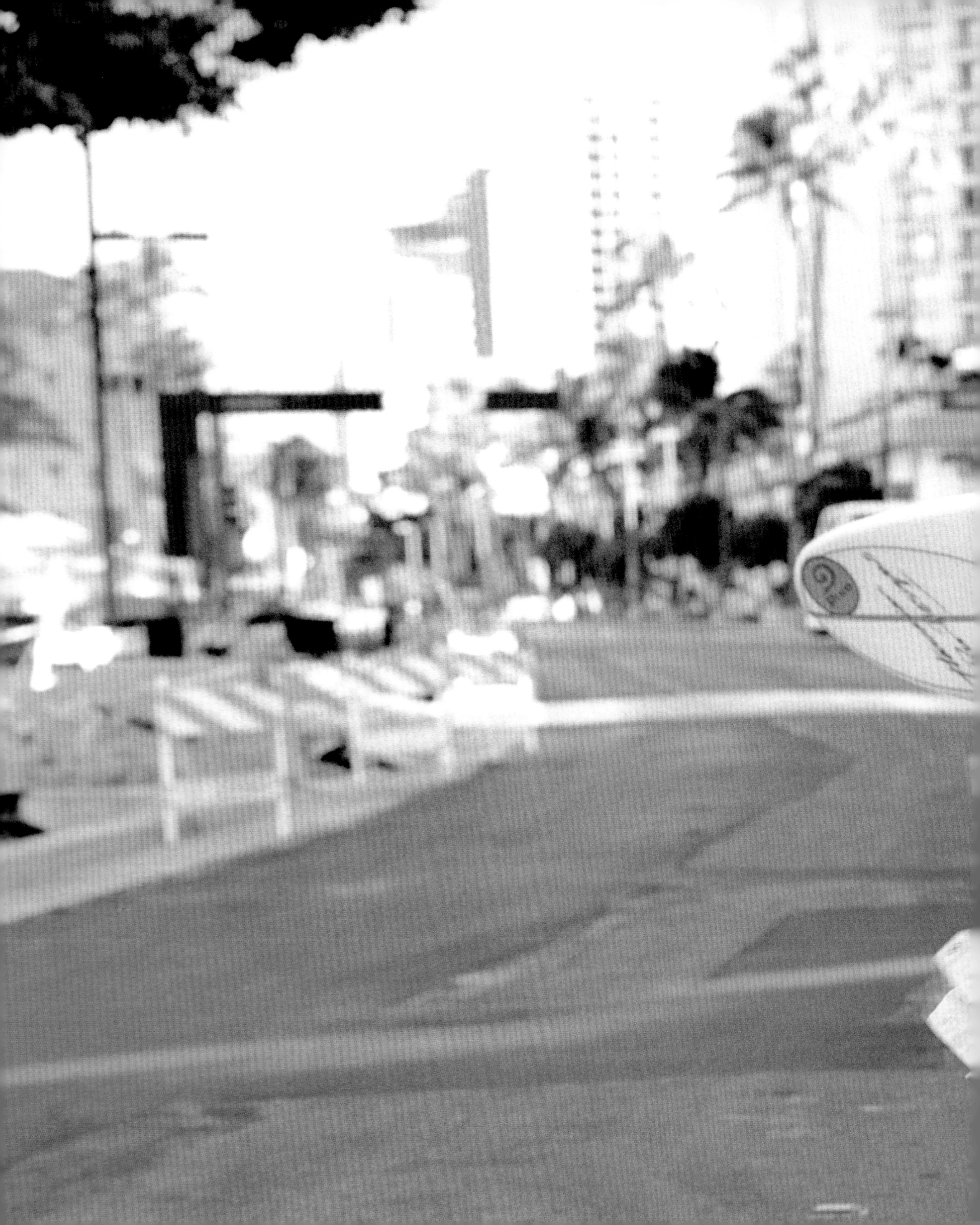

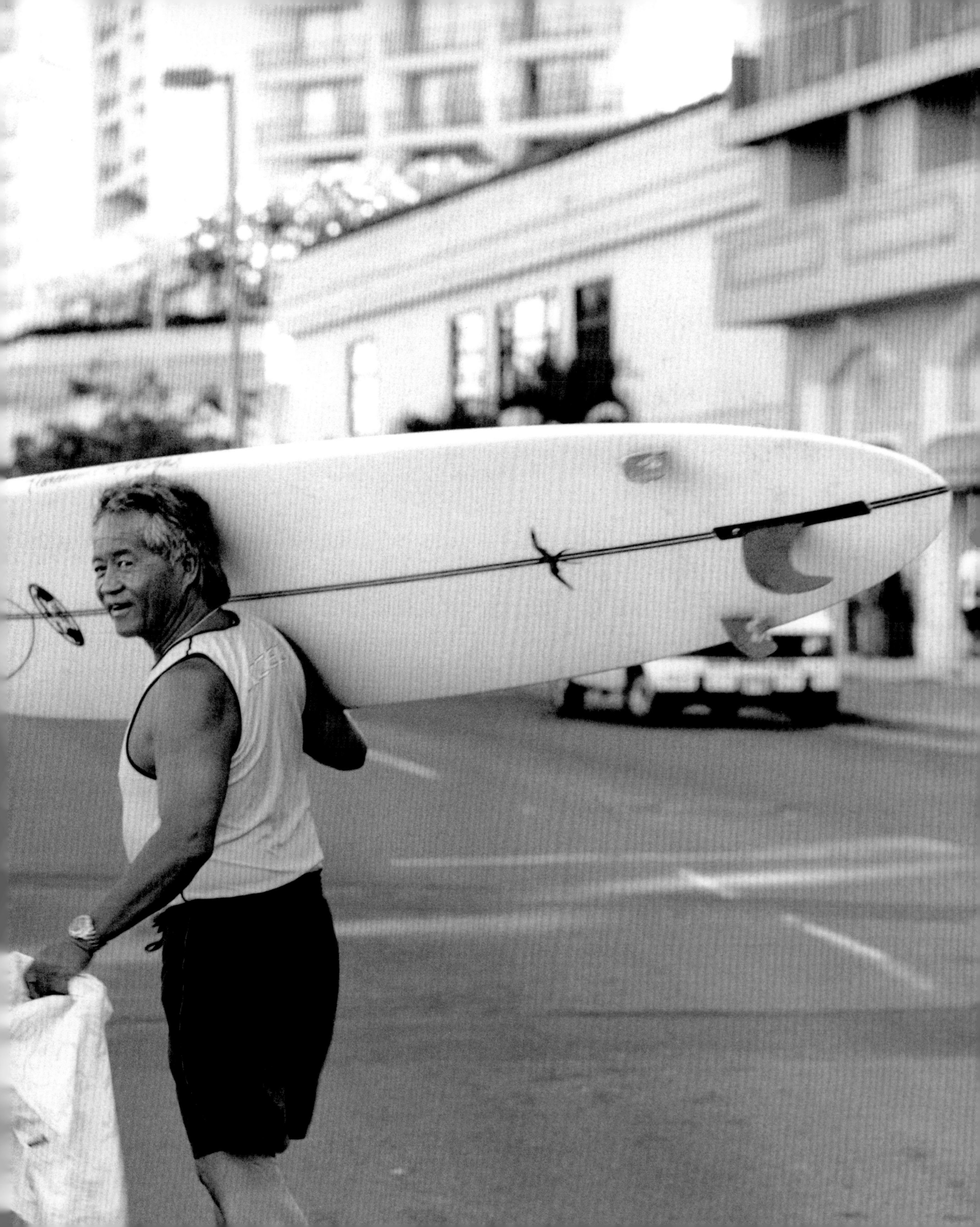

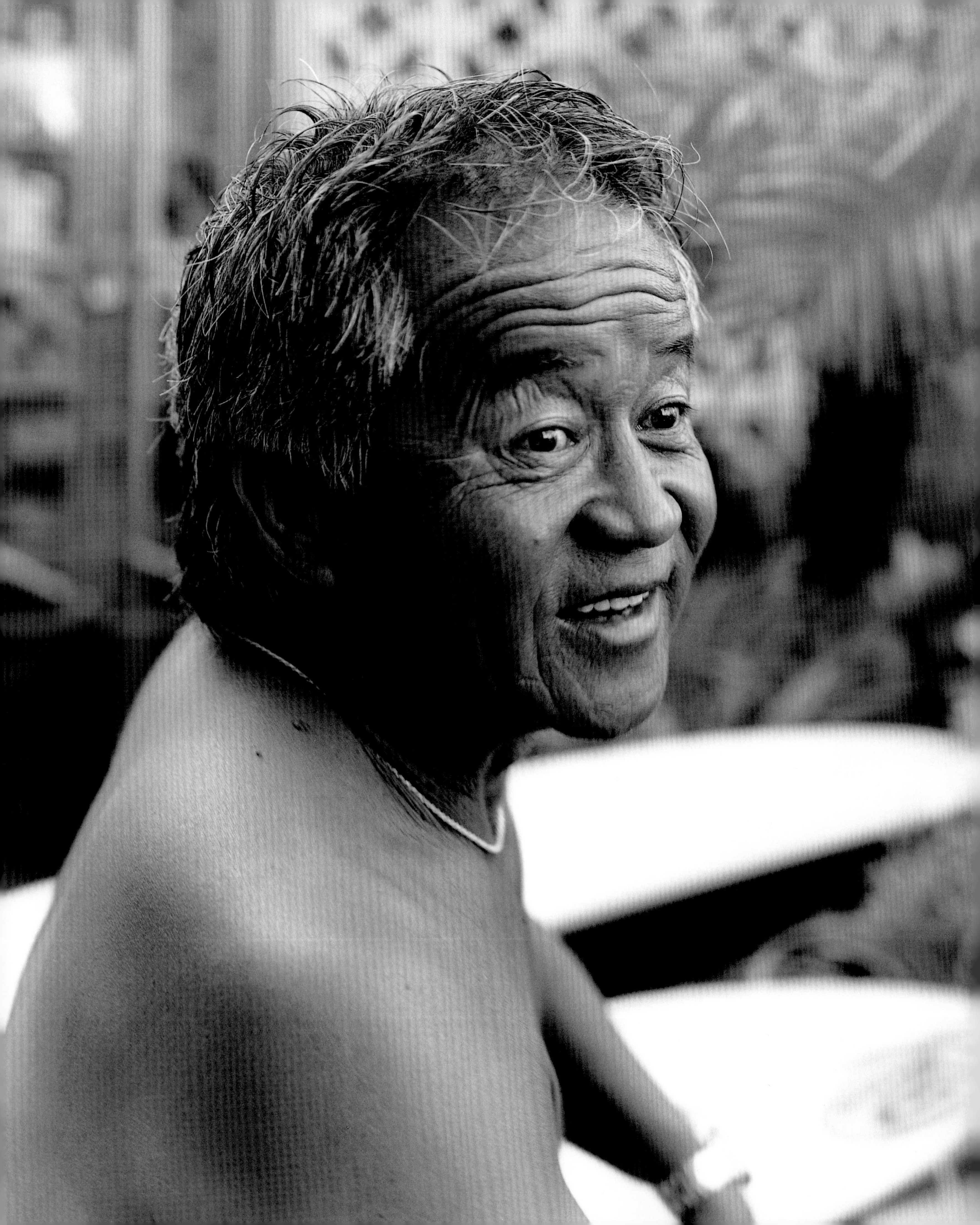

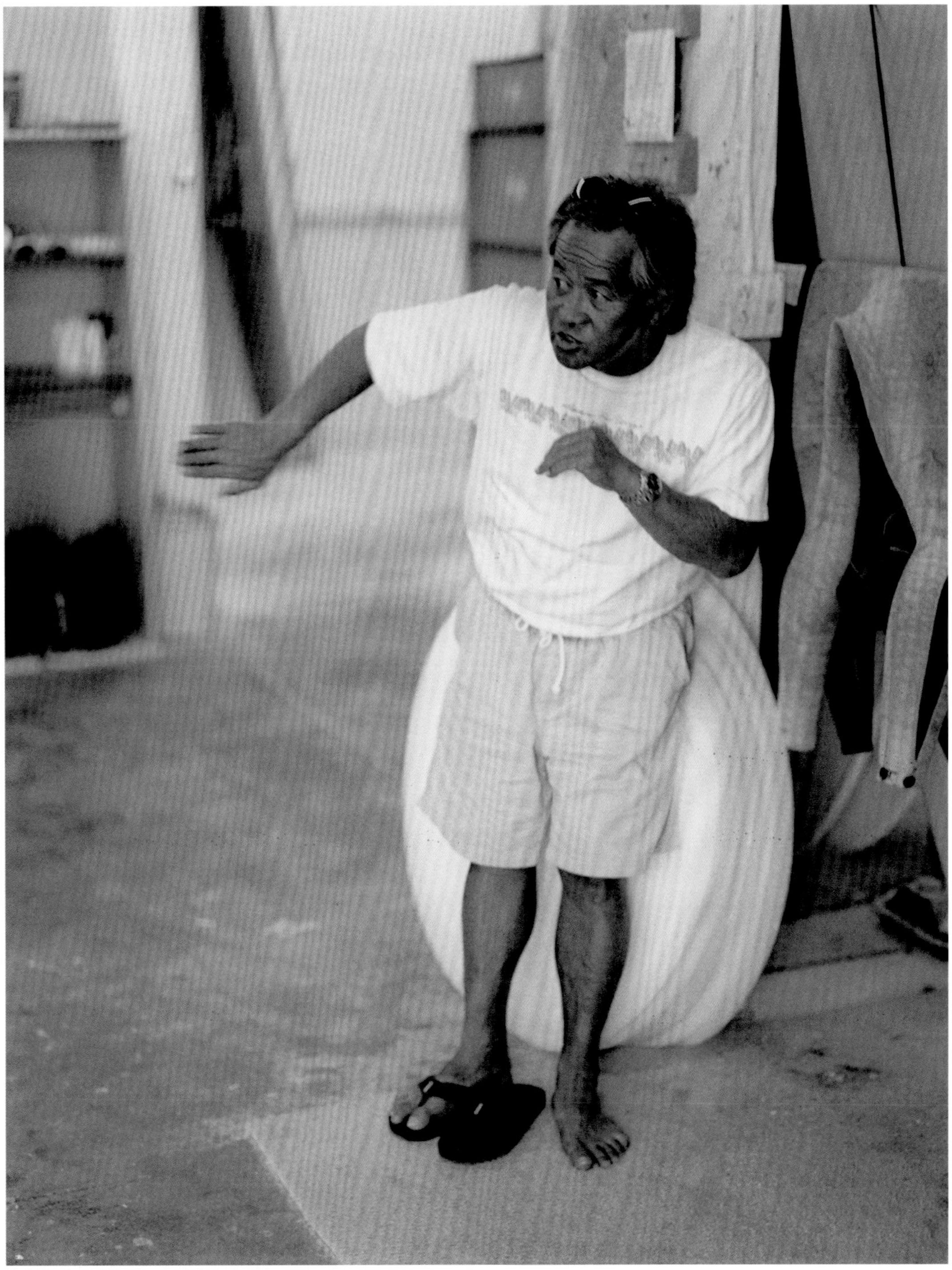

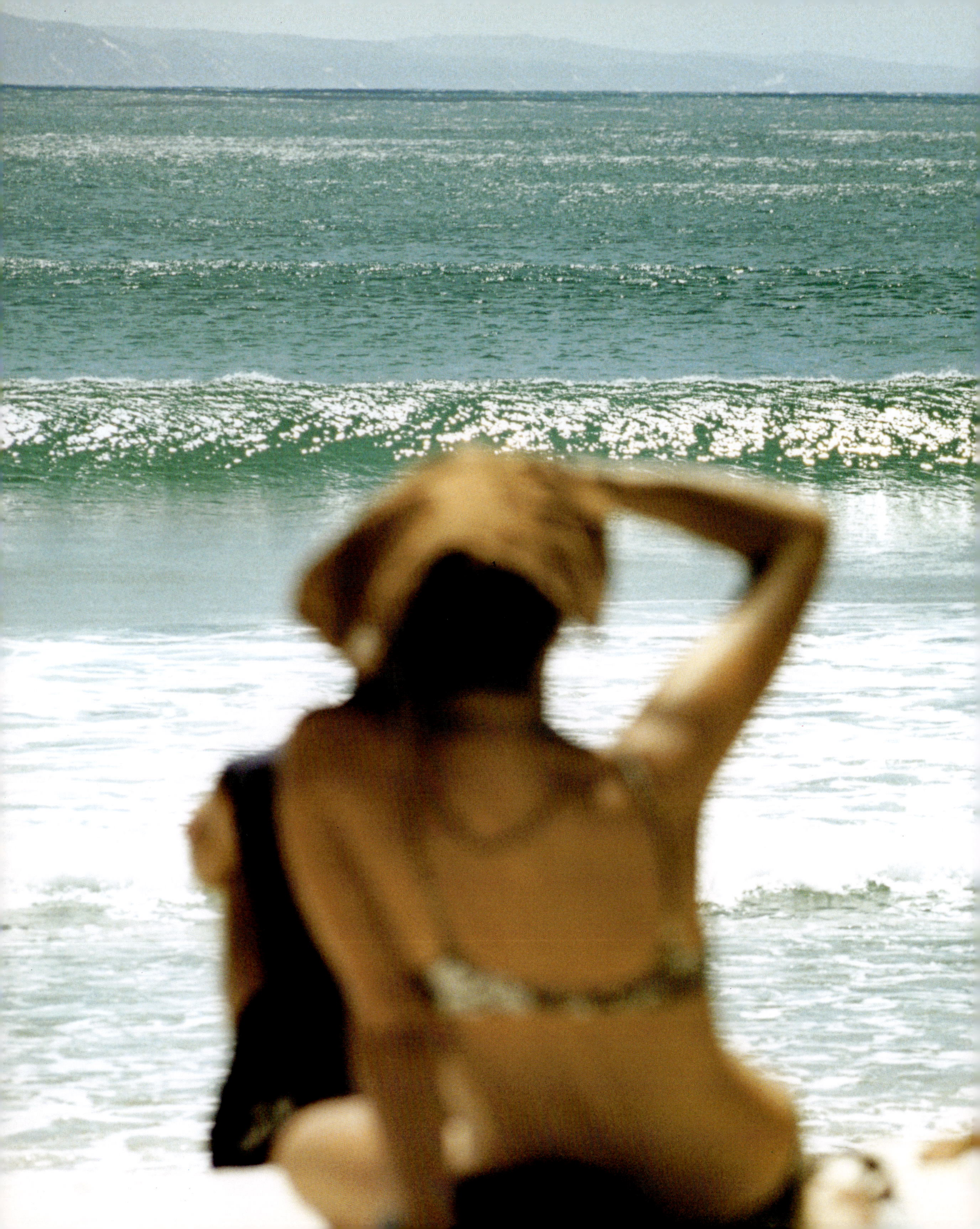

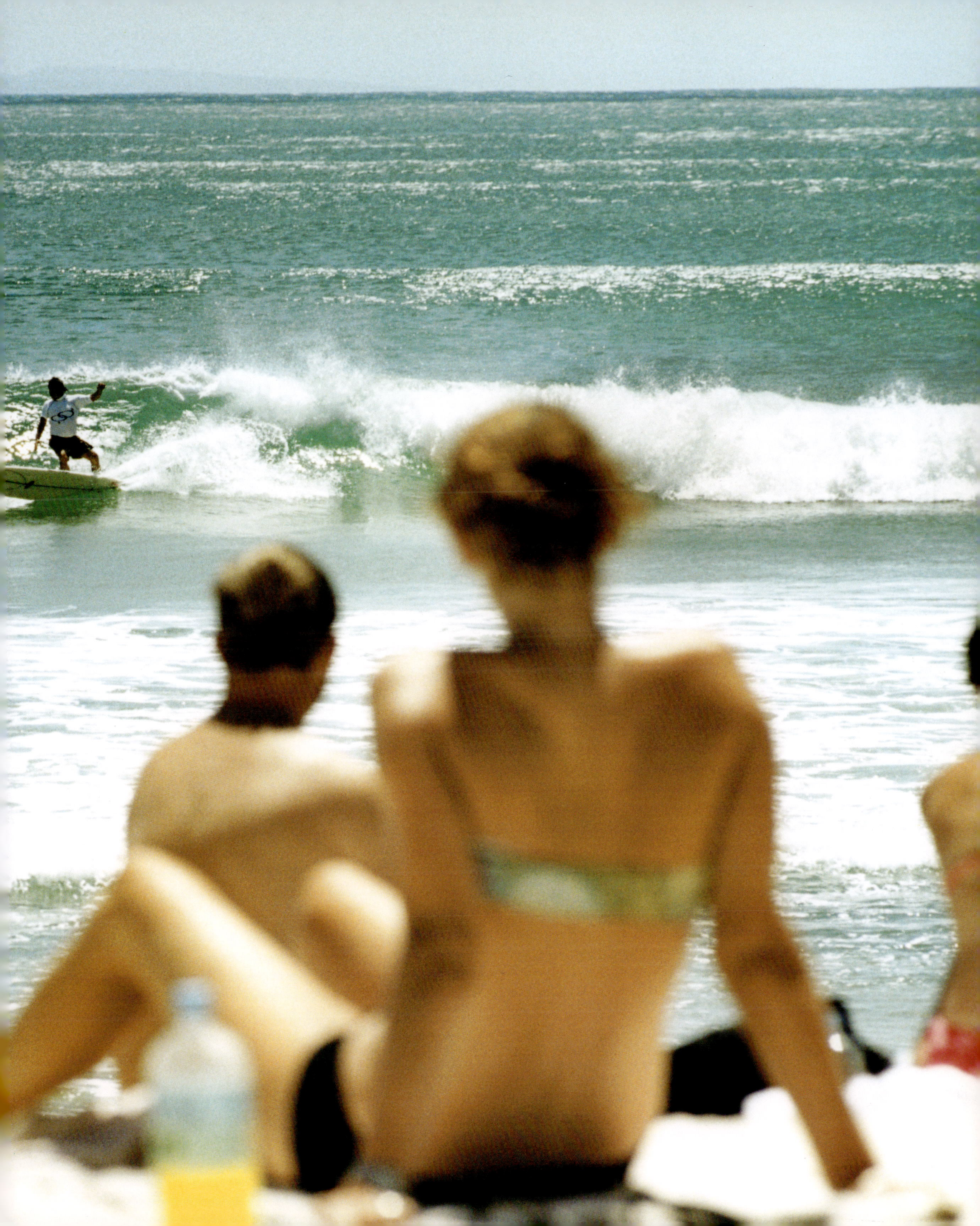

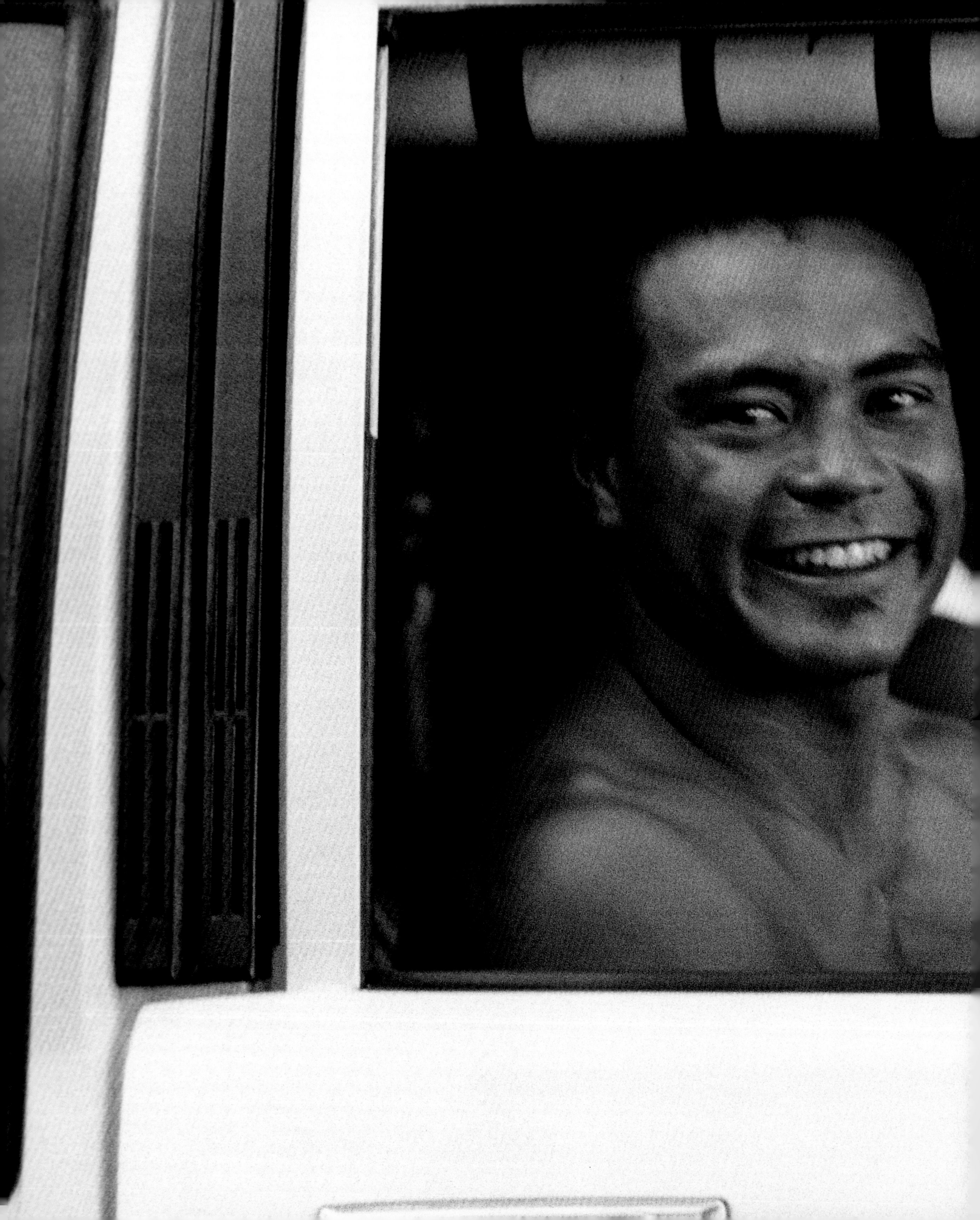

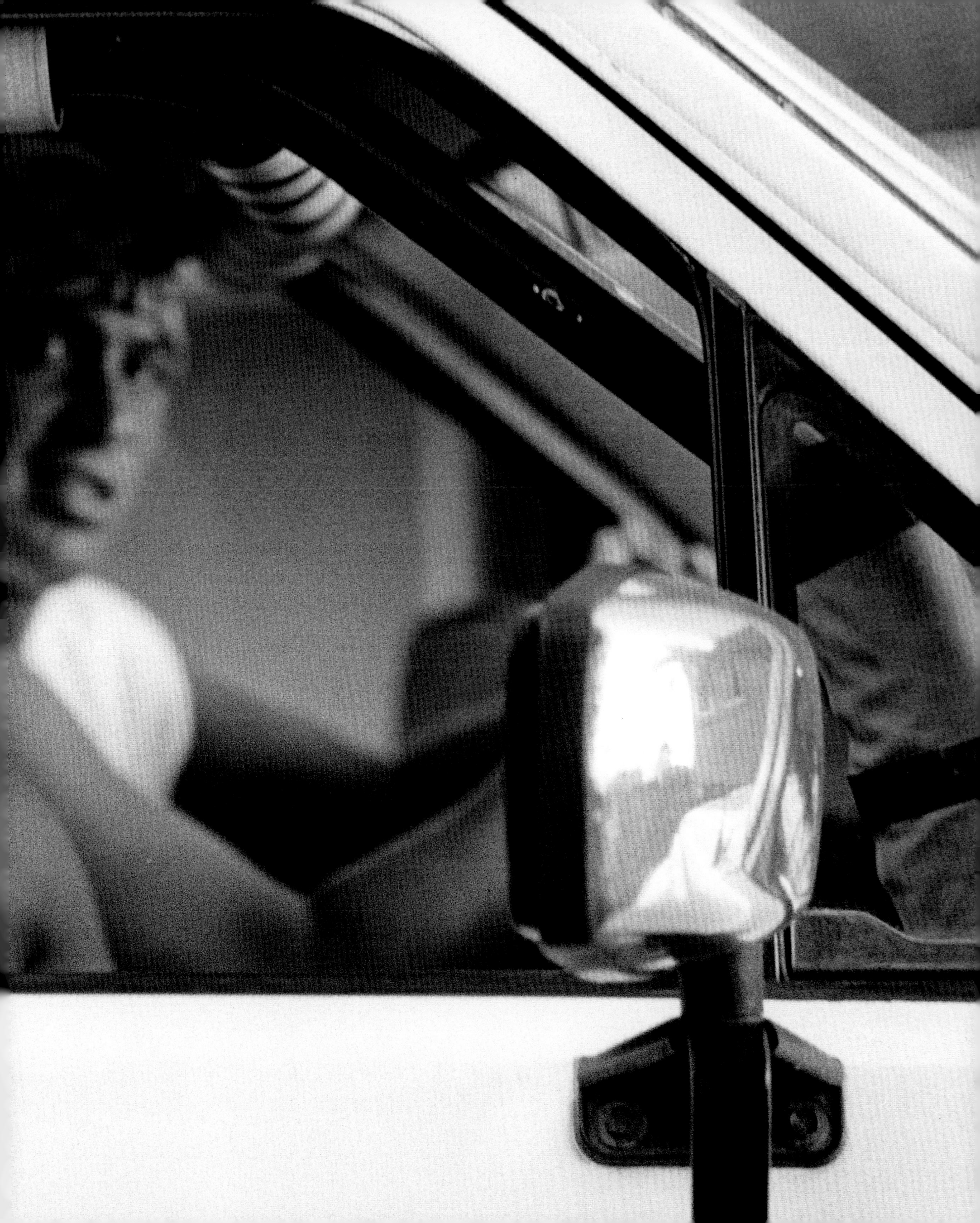

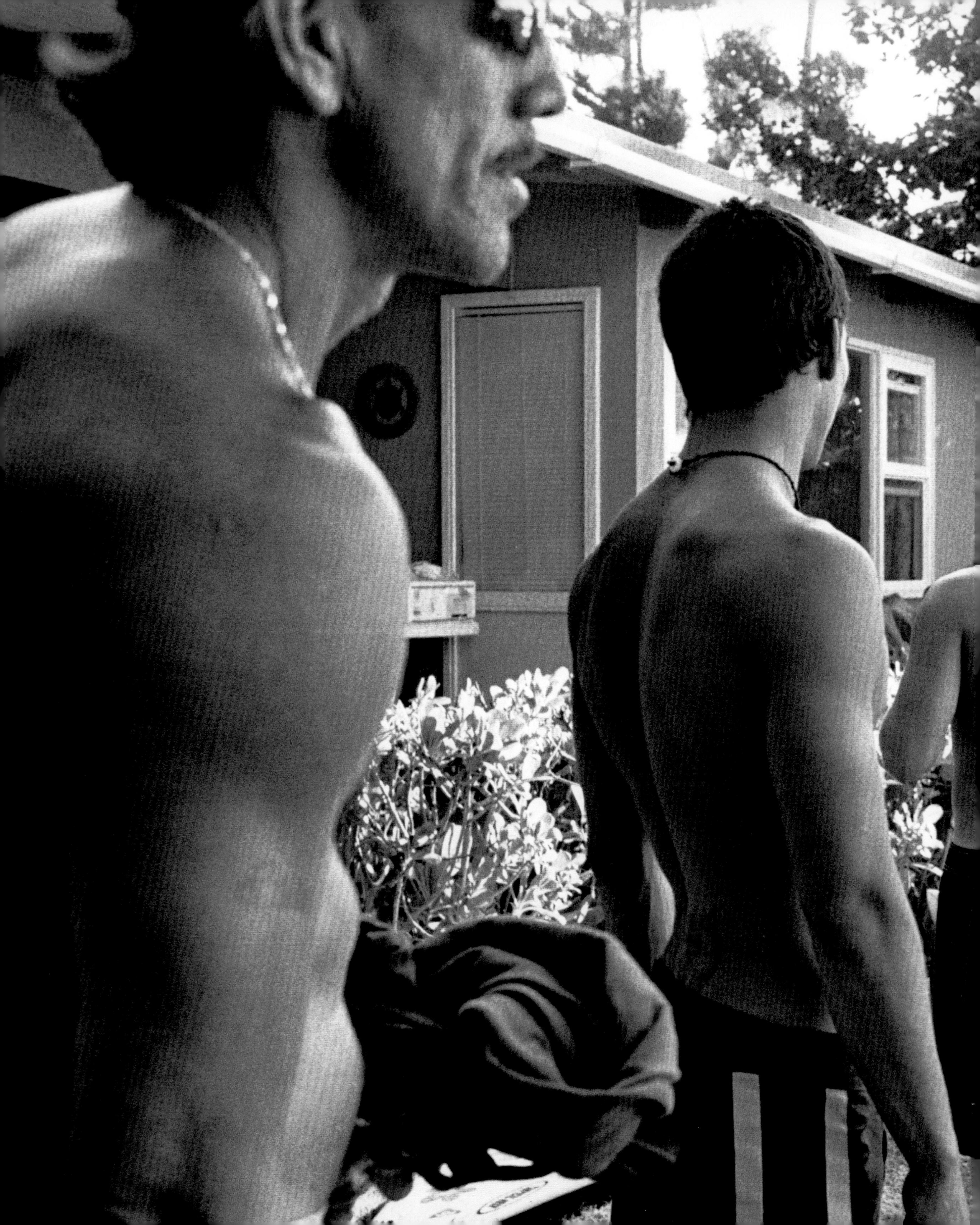

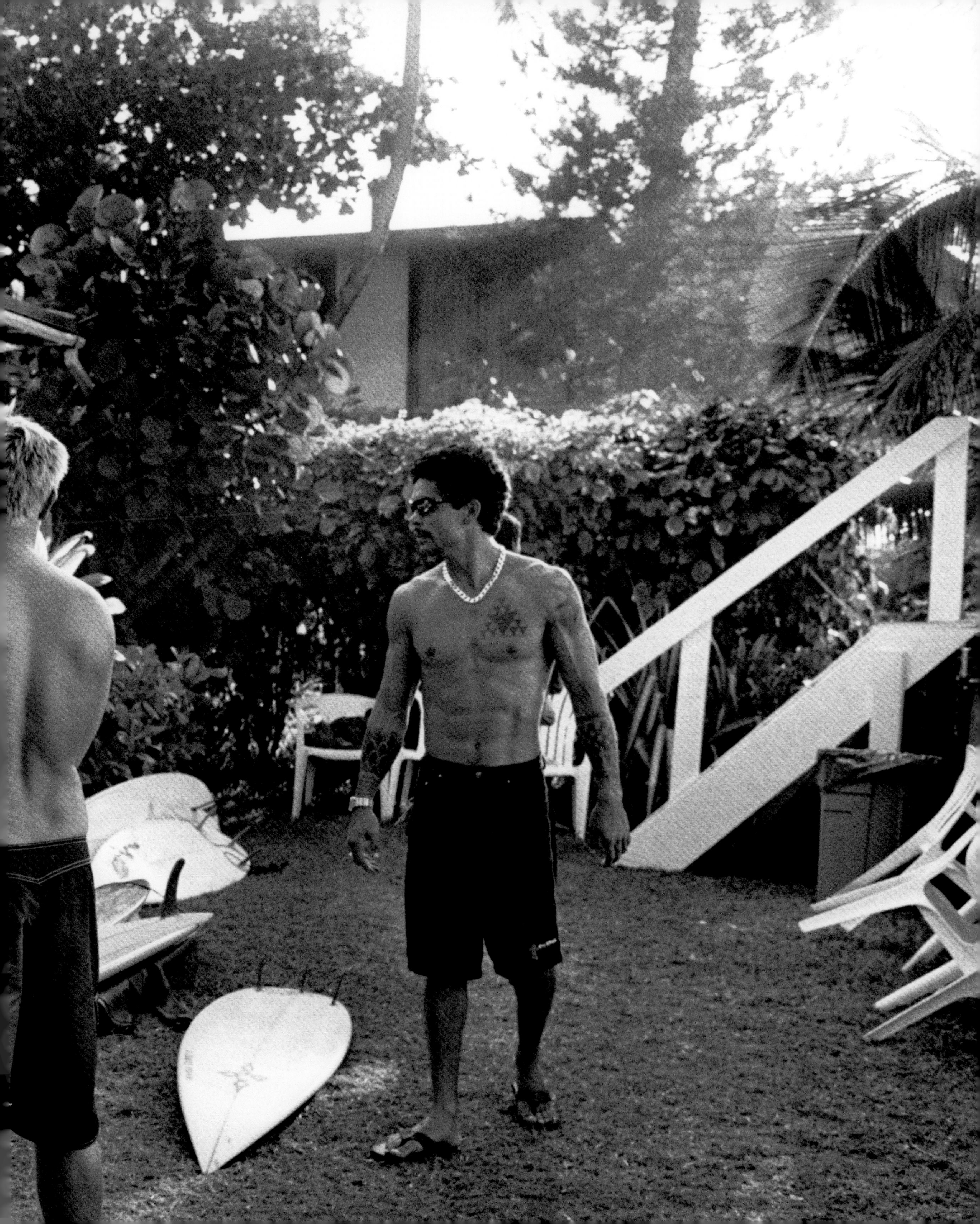

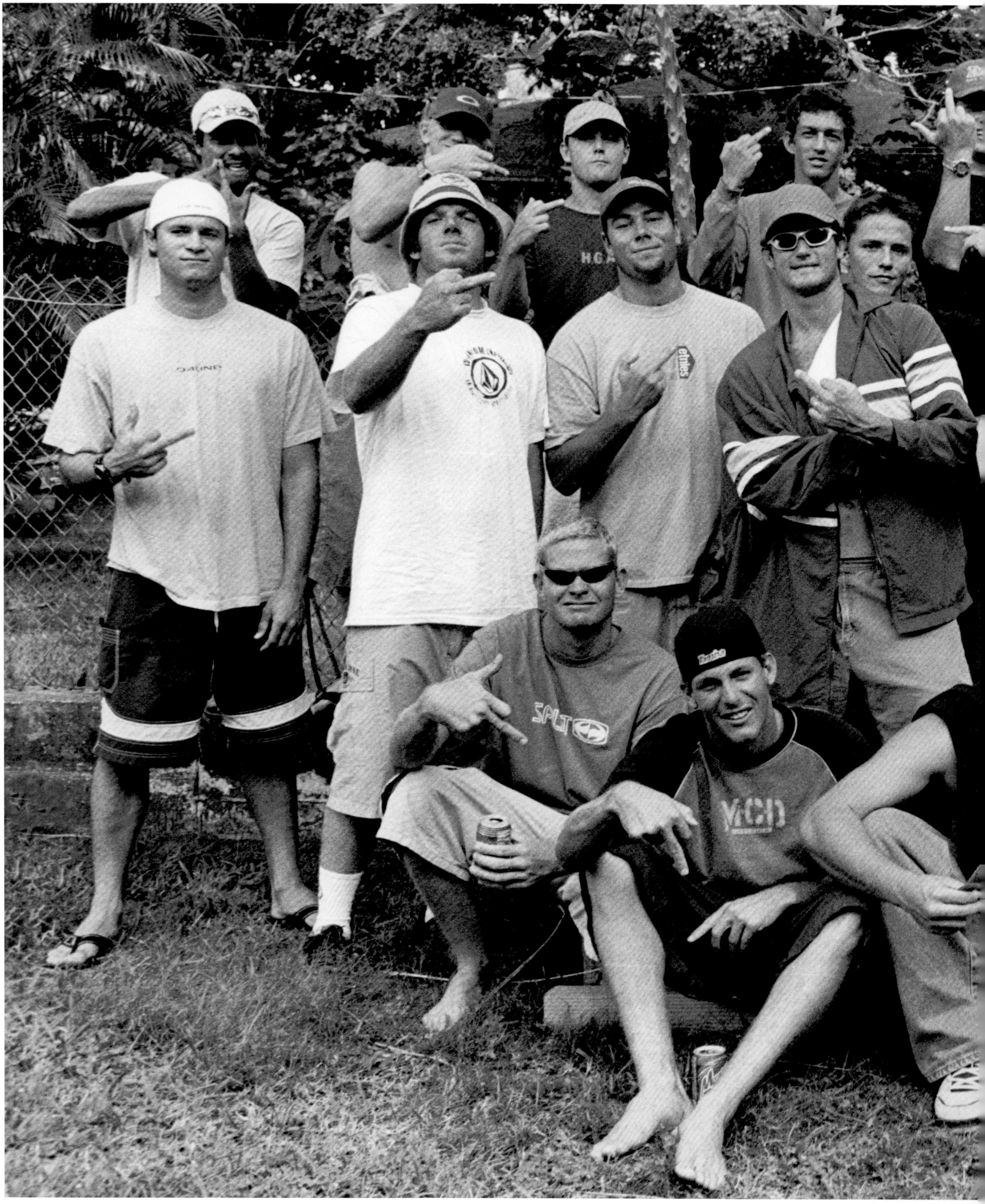
etnies
SPLIT
MCD

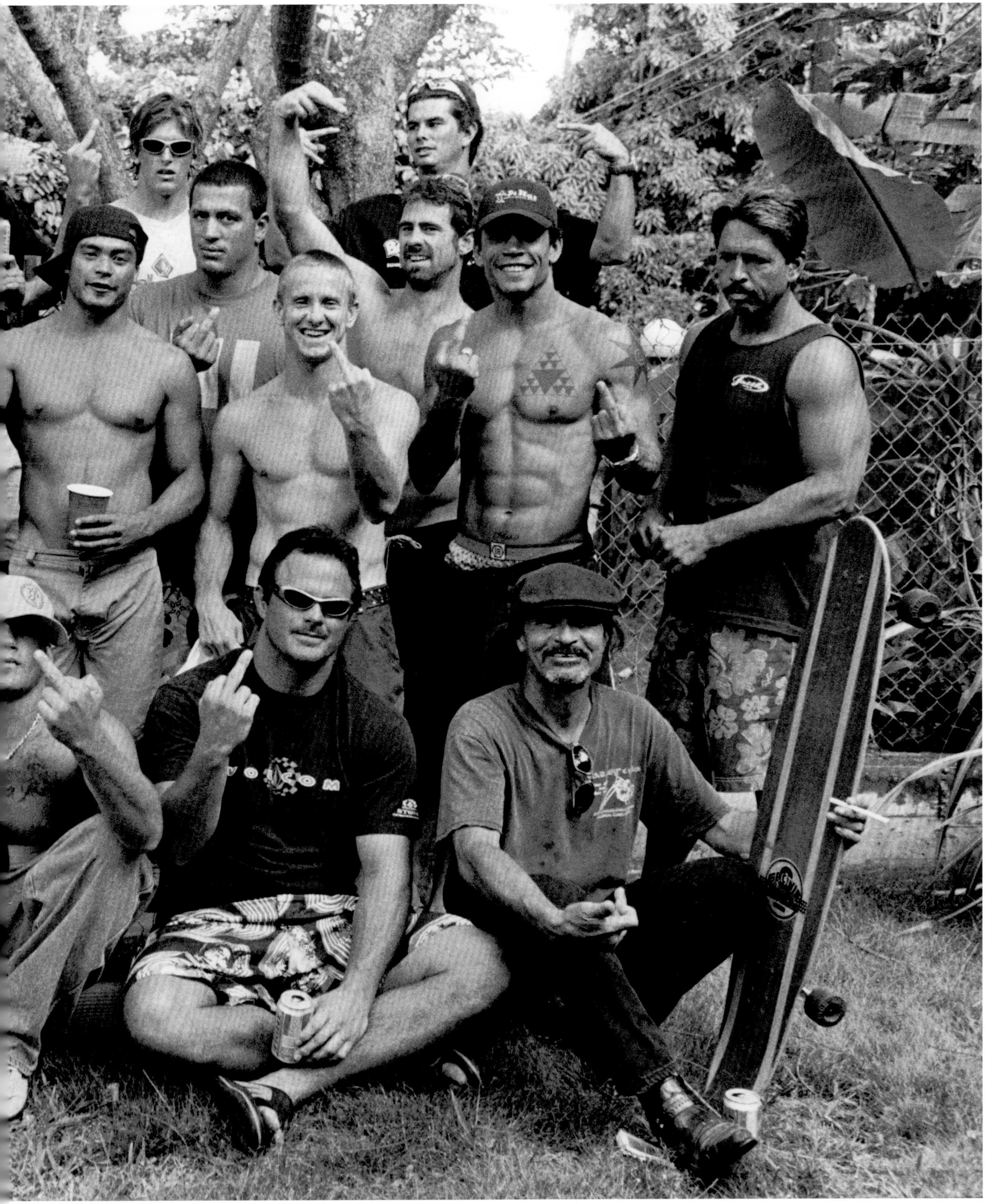

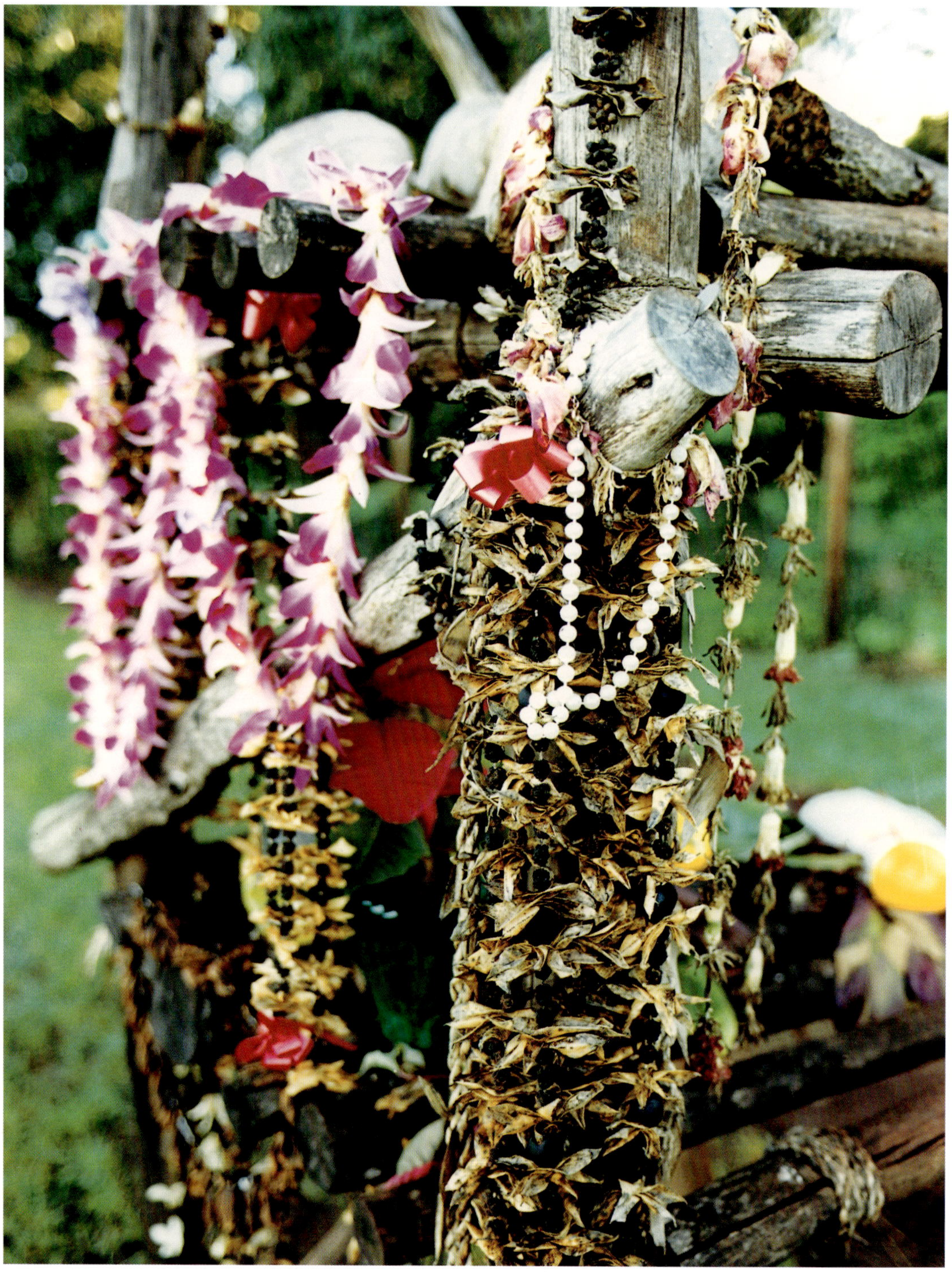

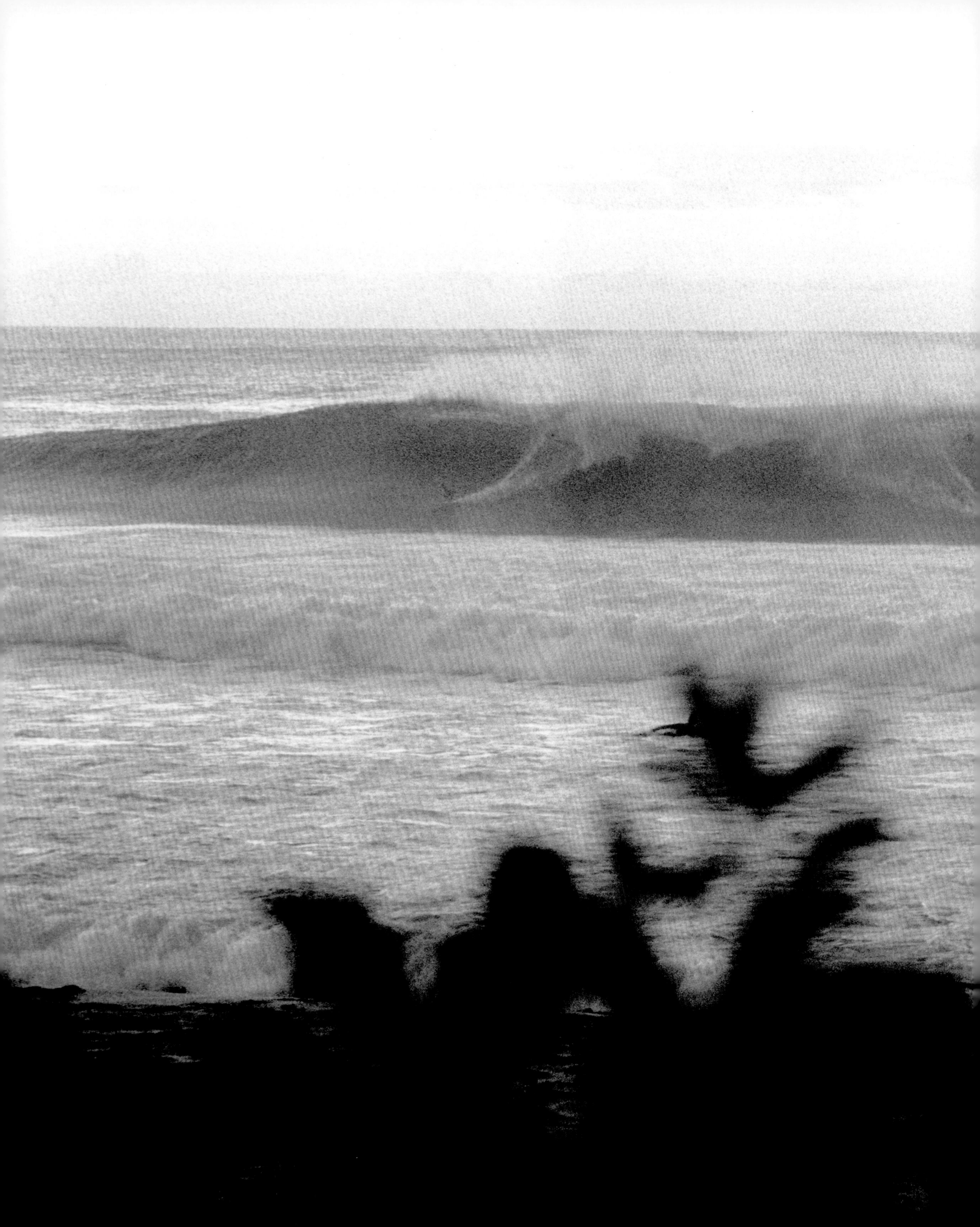

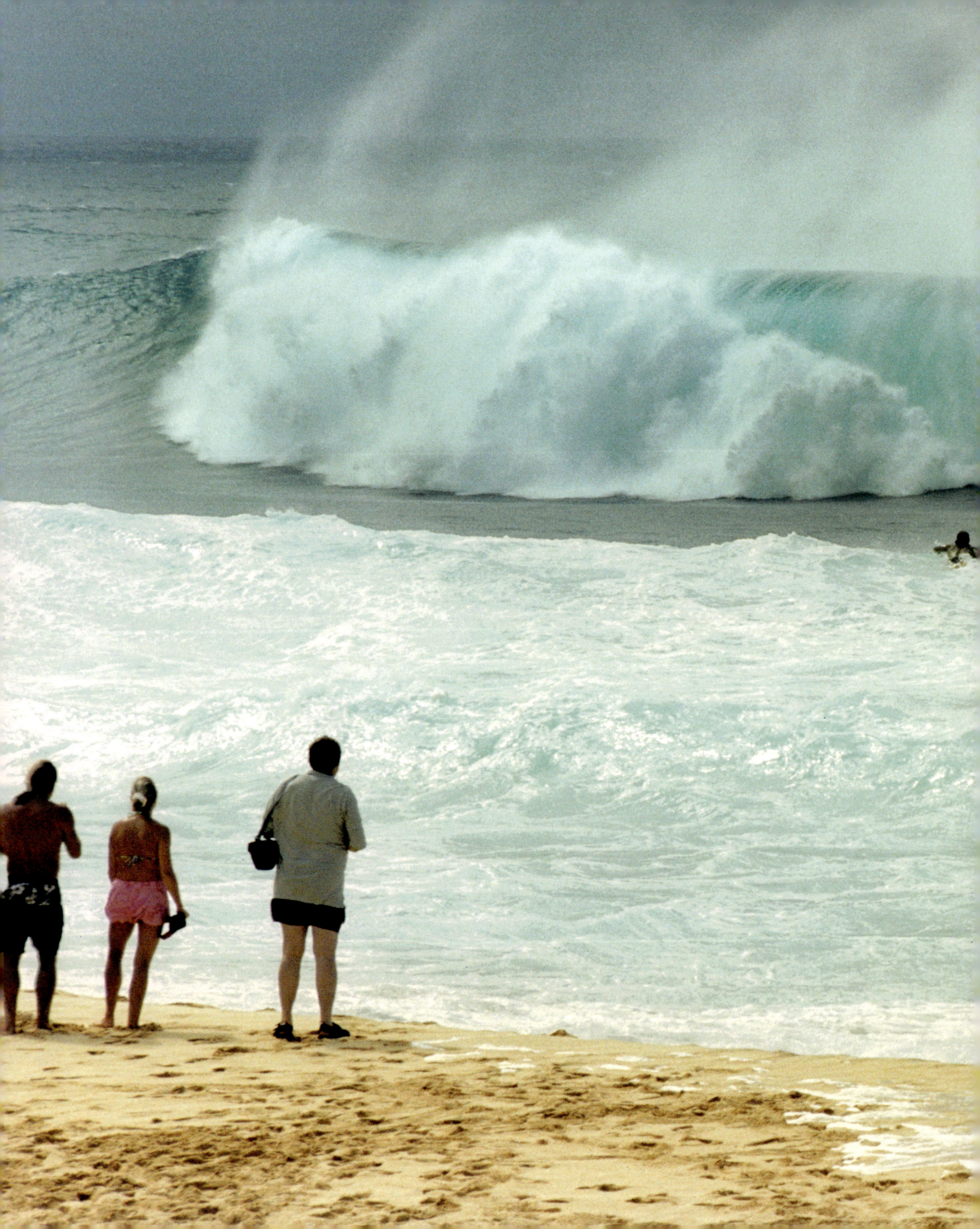

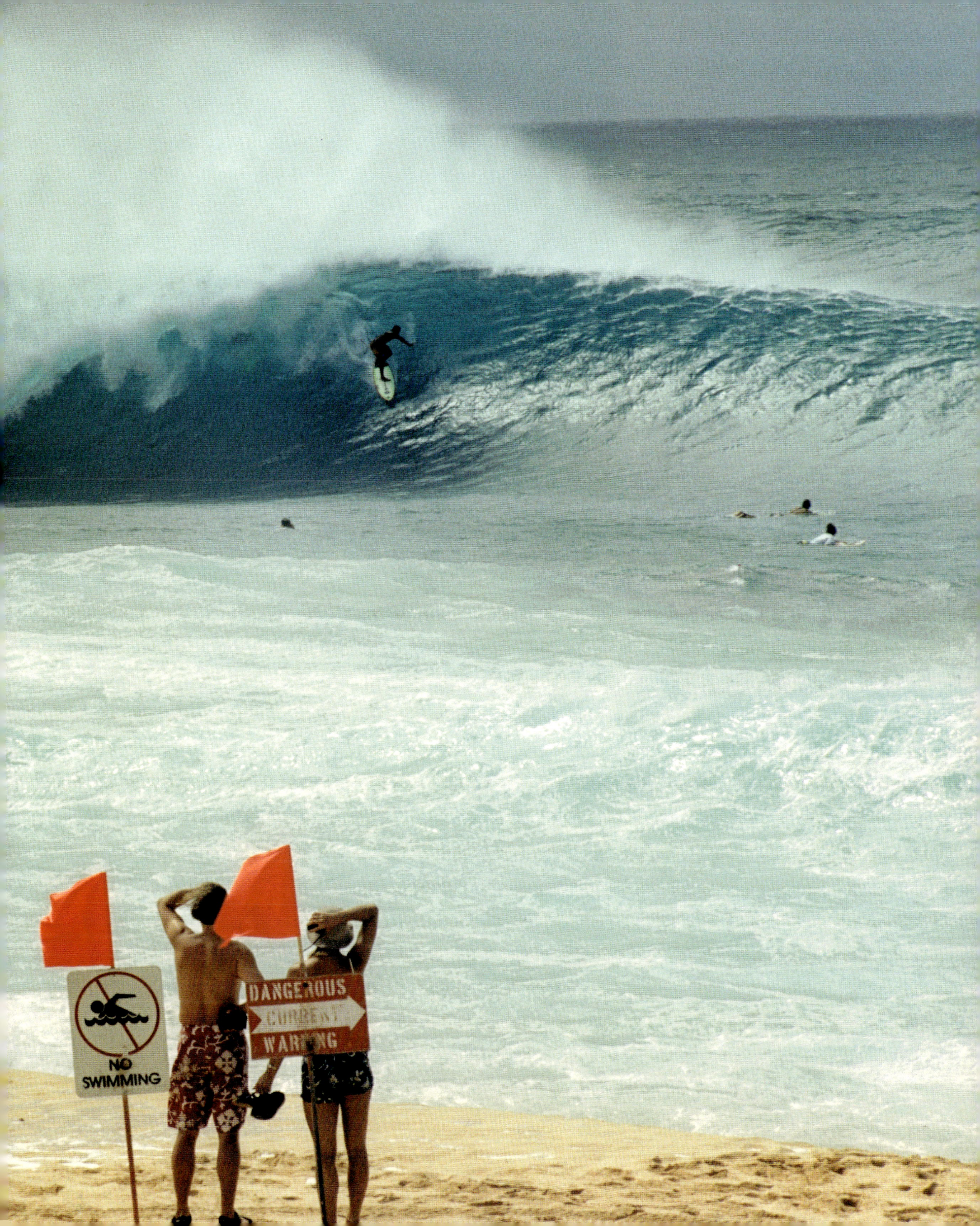
NO SWIMMING
DANGEROUS
CURRENT
WARNING

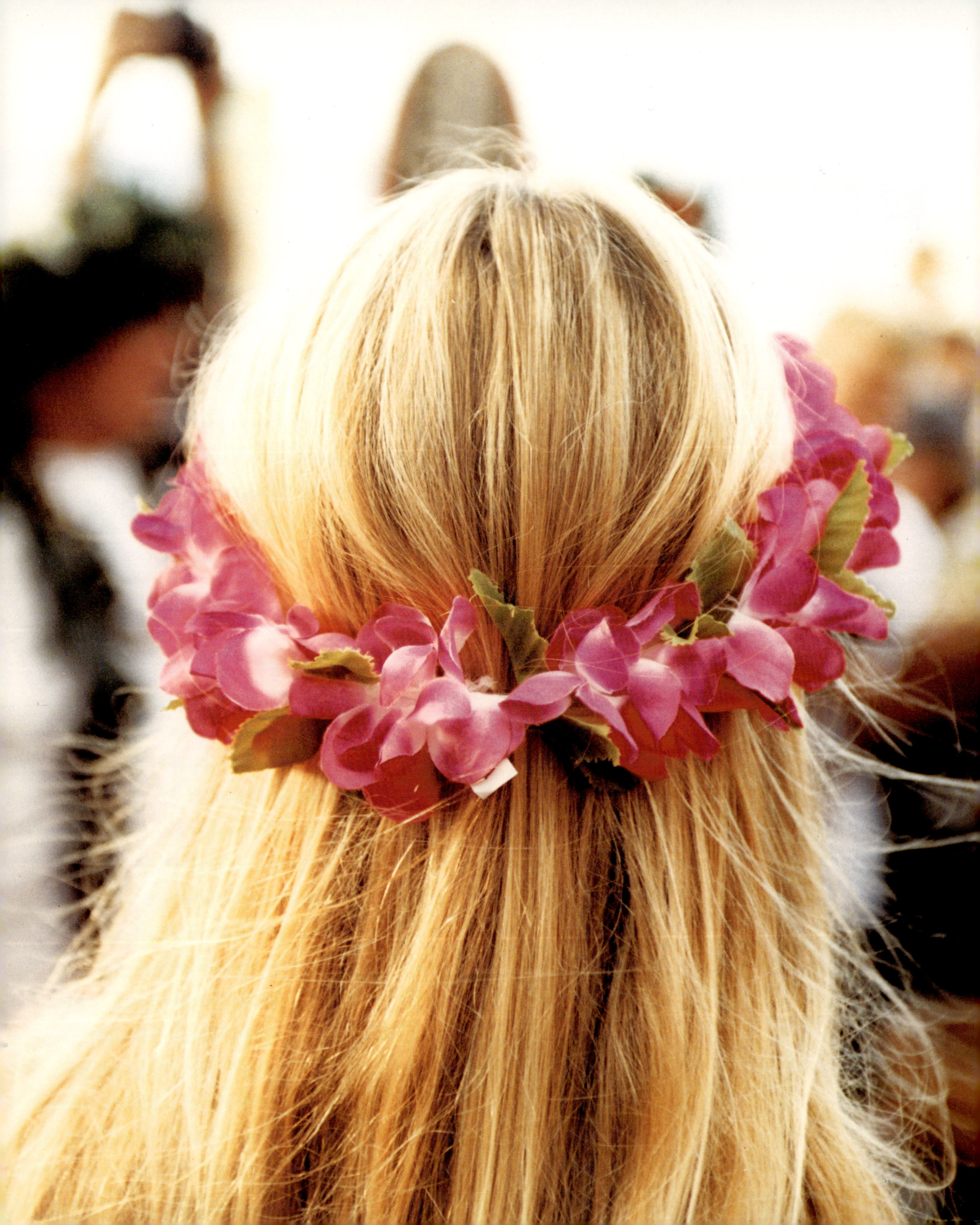

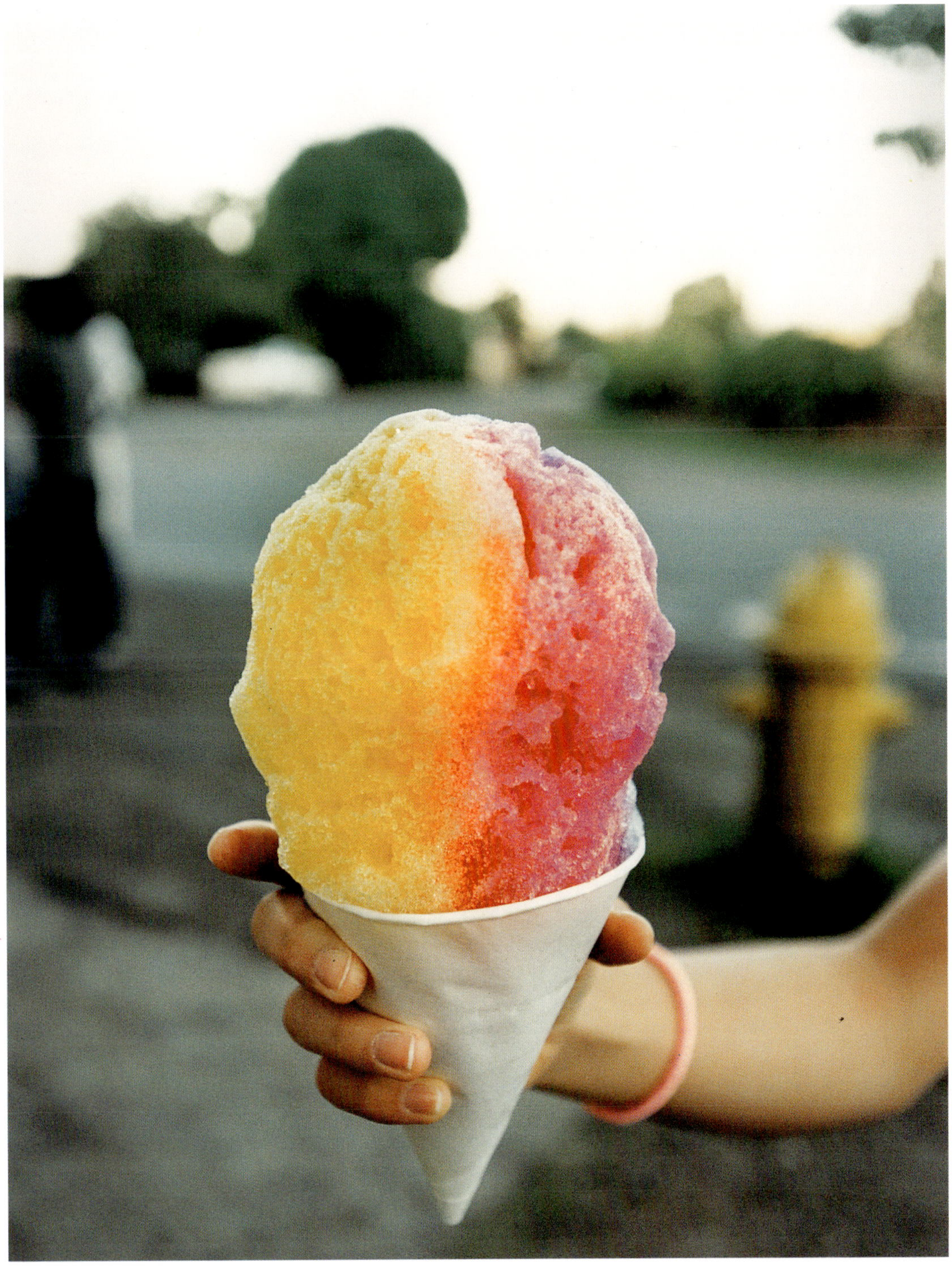

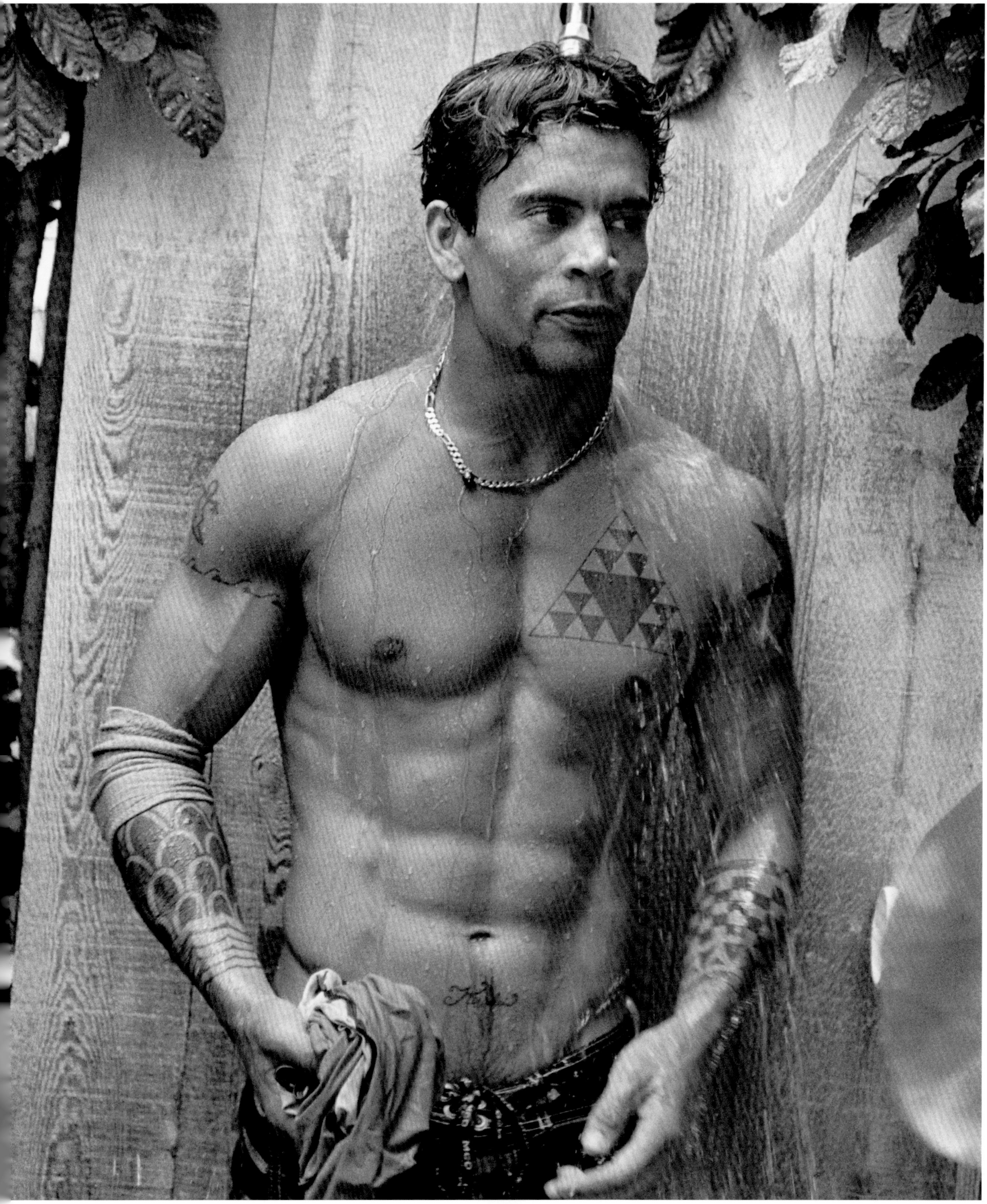

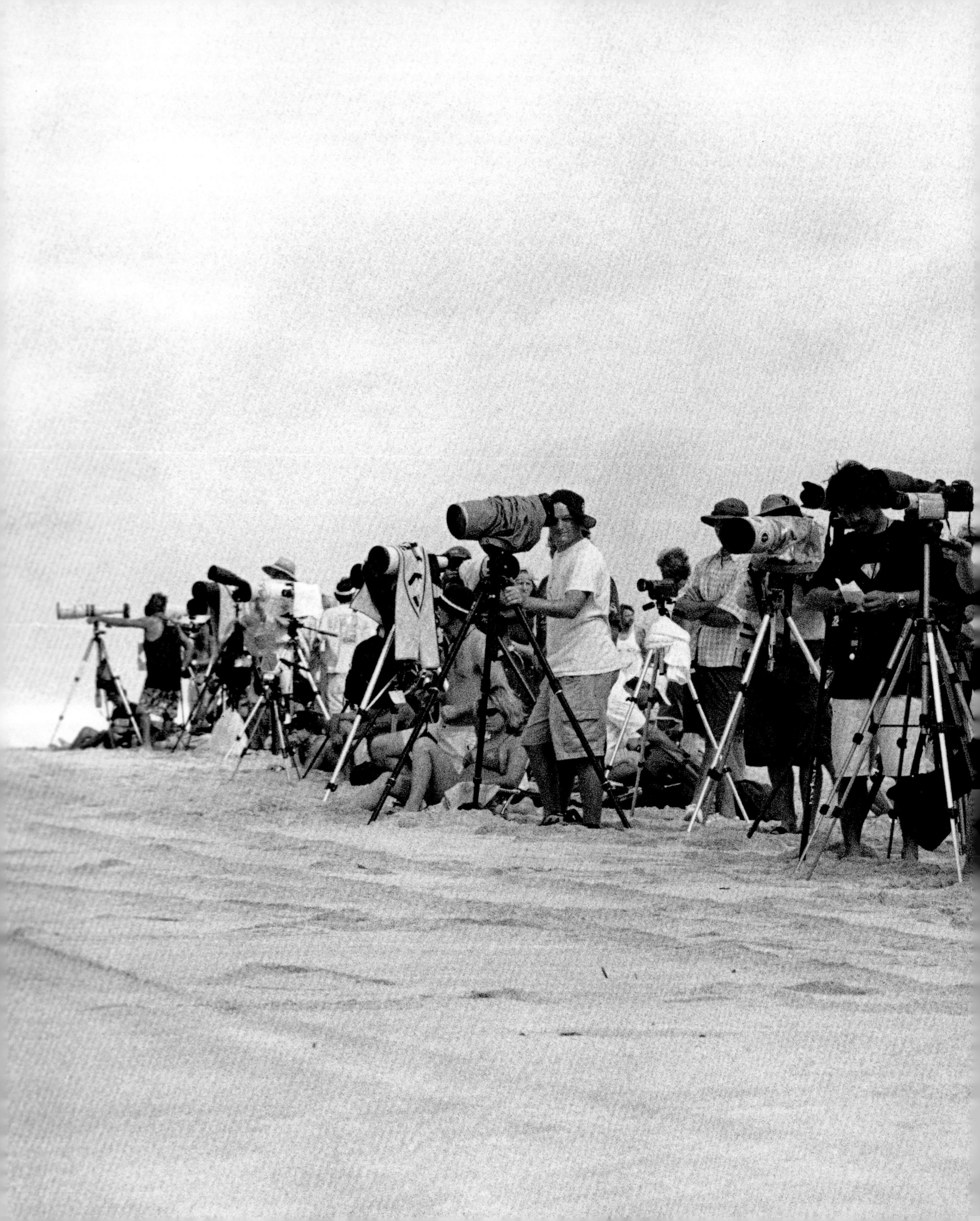

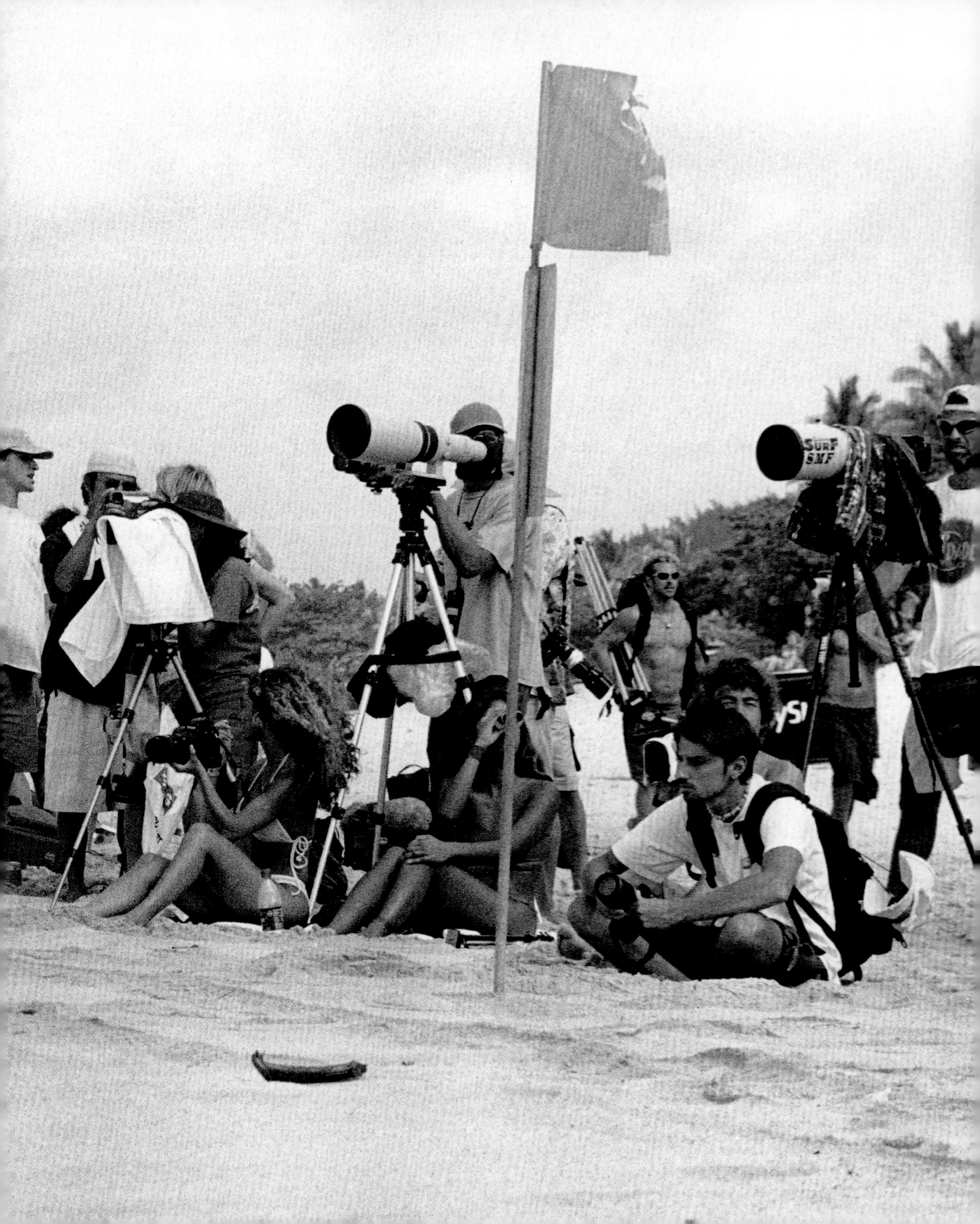
SURF
SMF

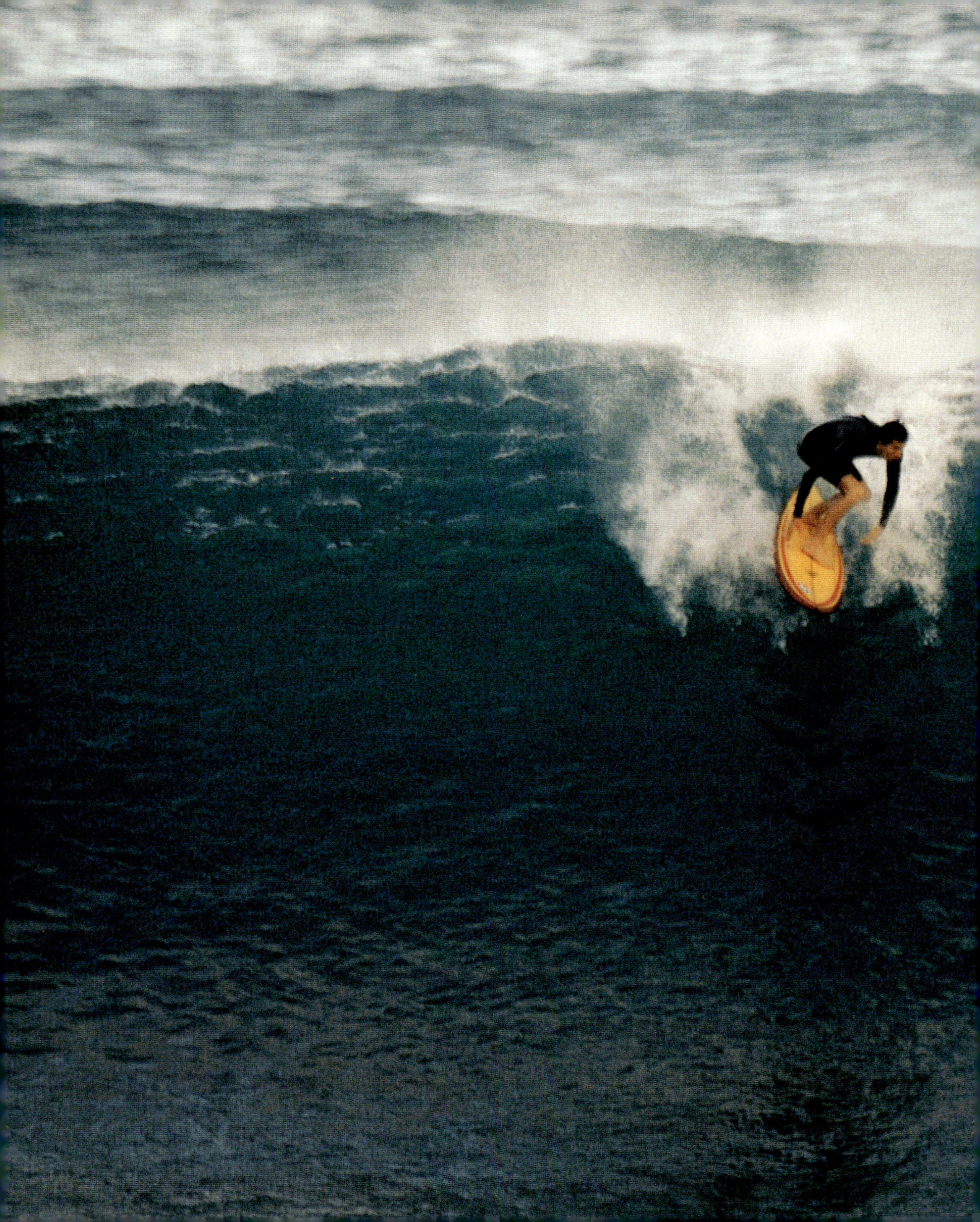

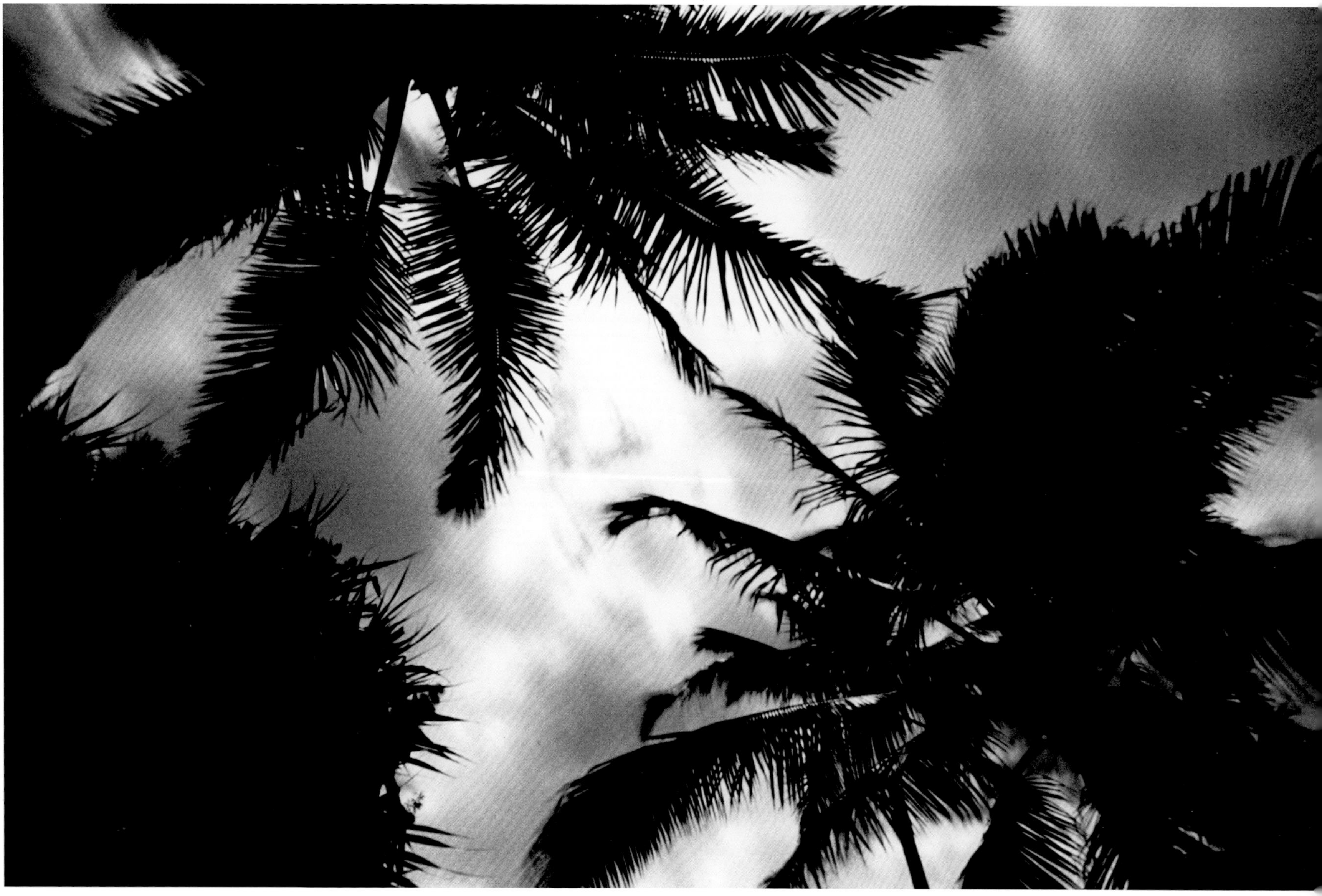

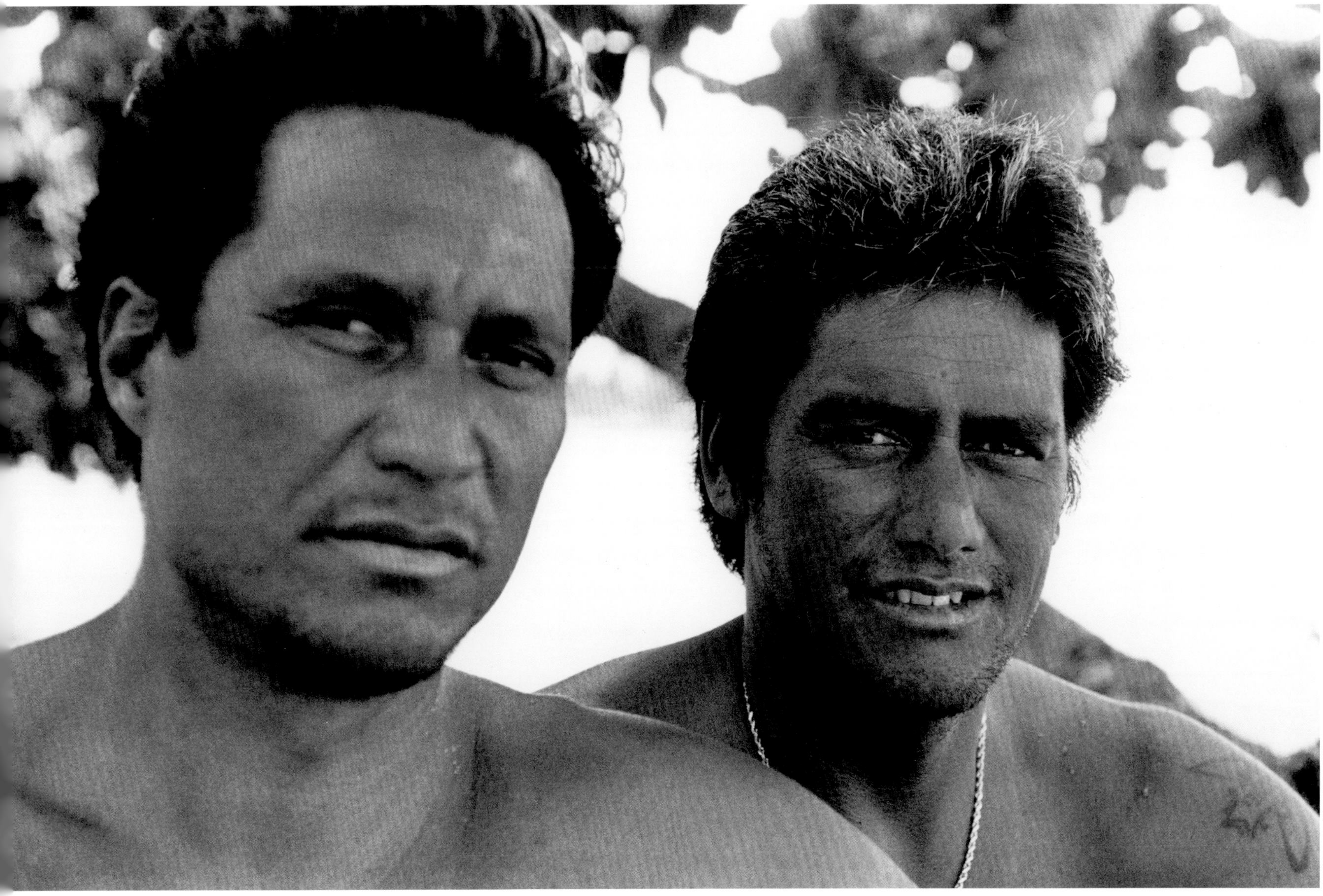

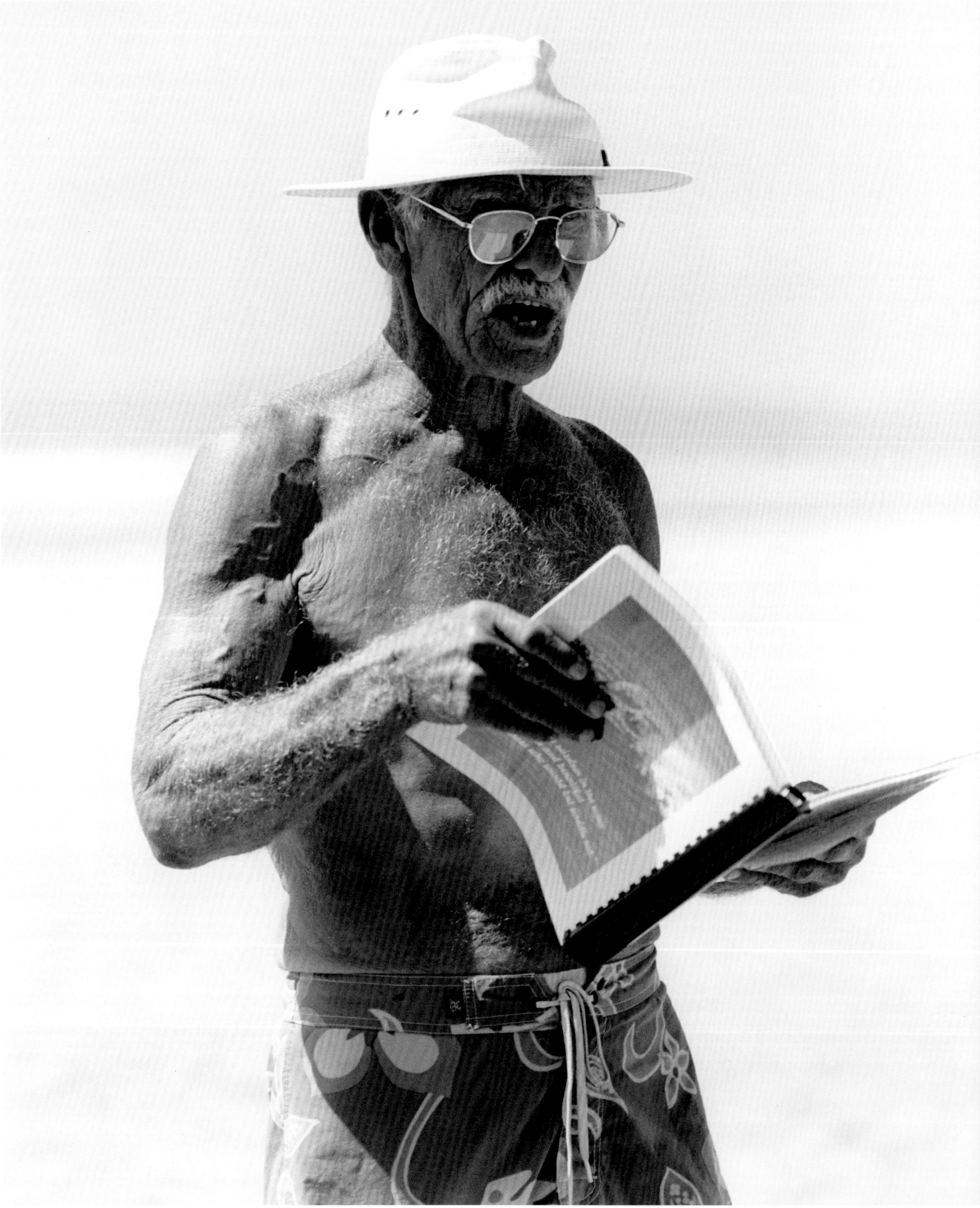

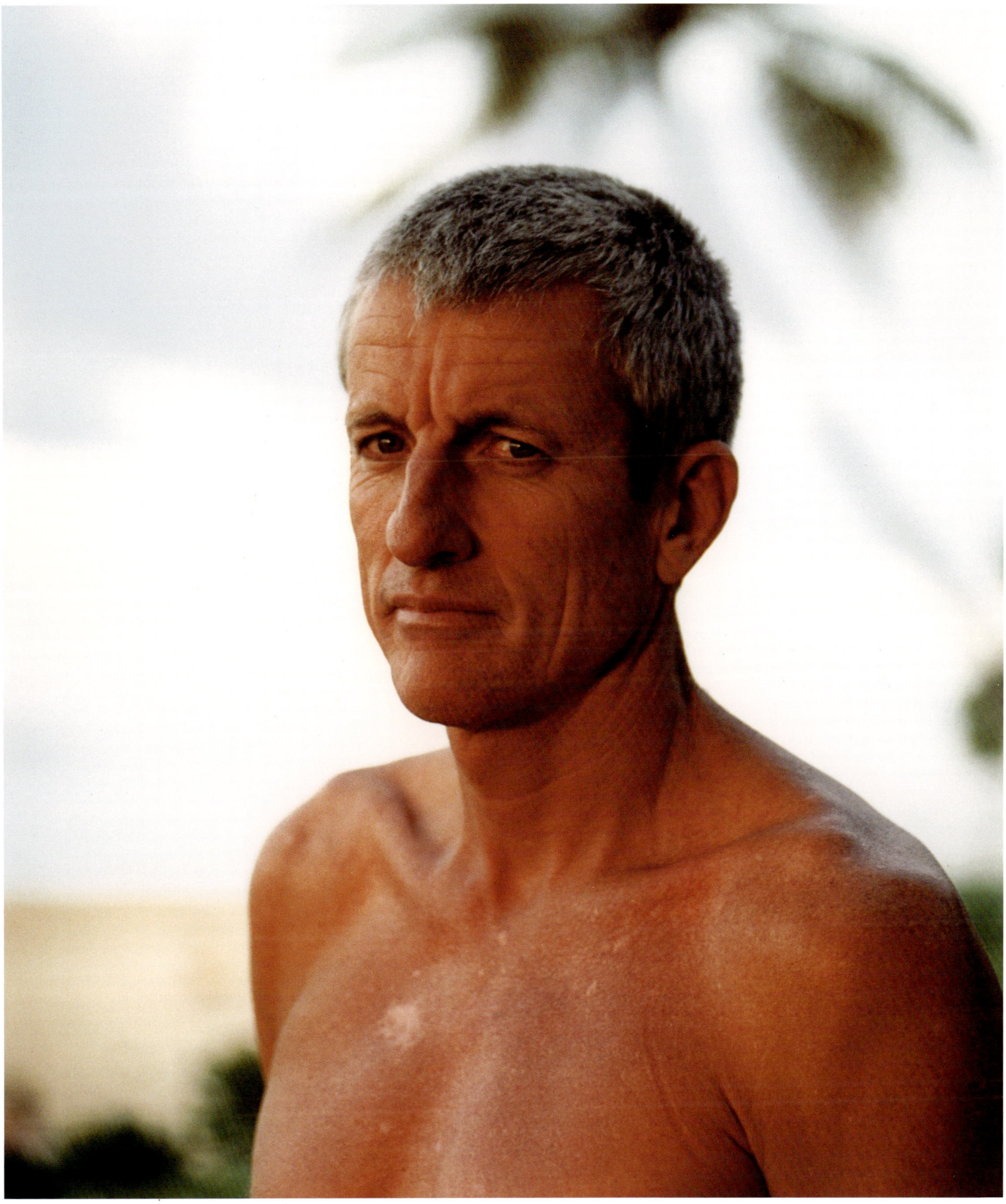

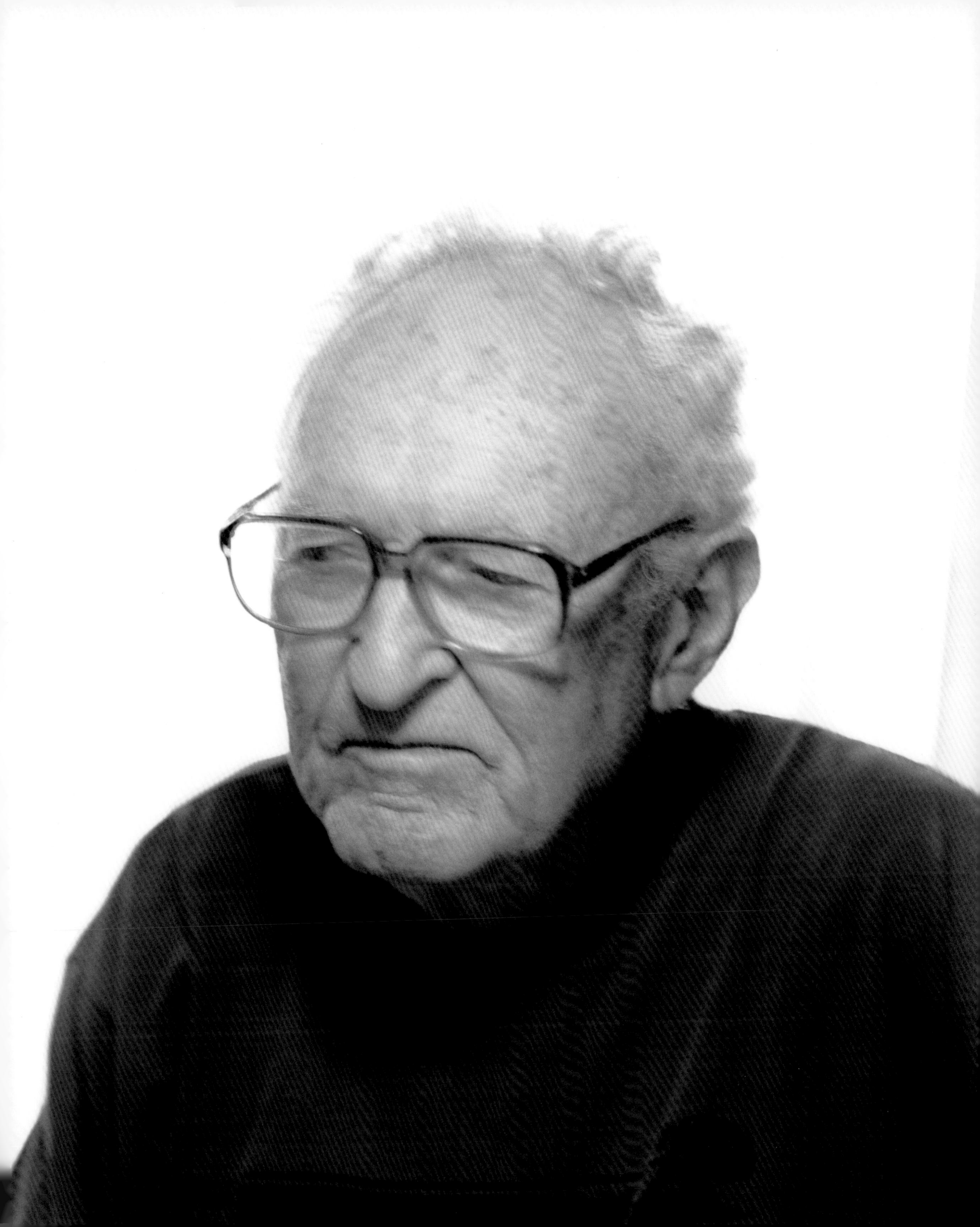

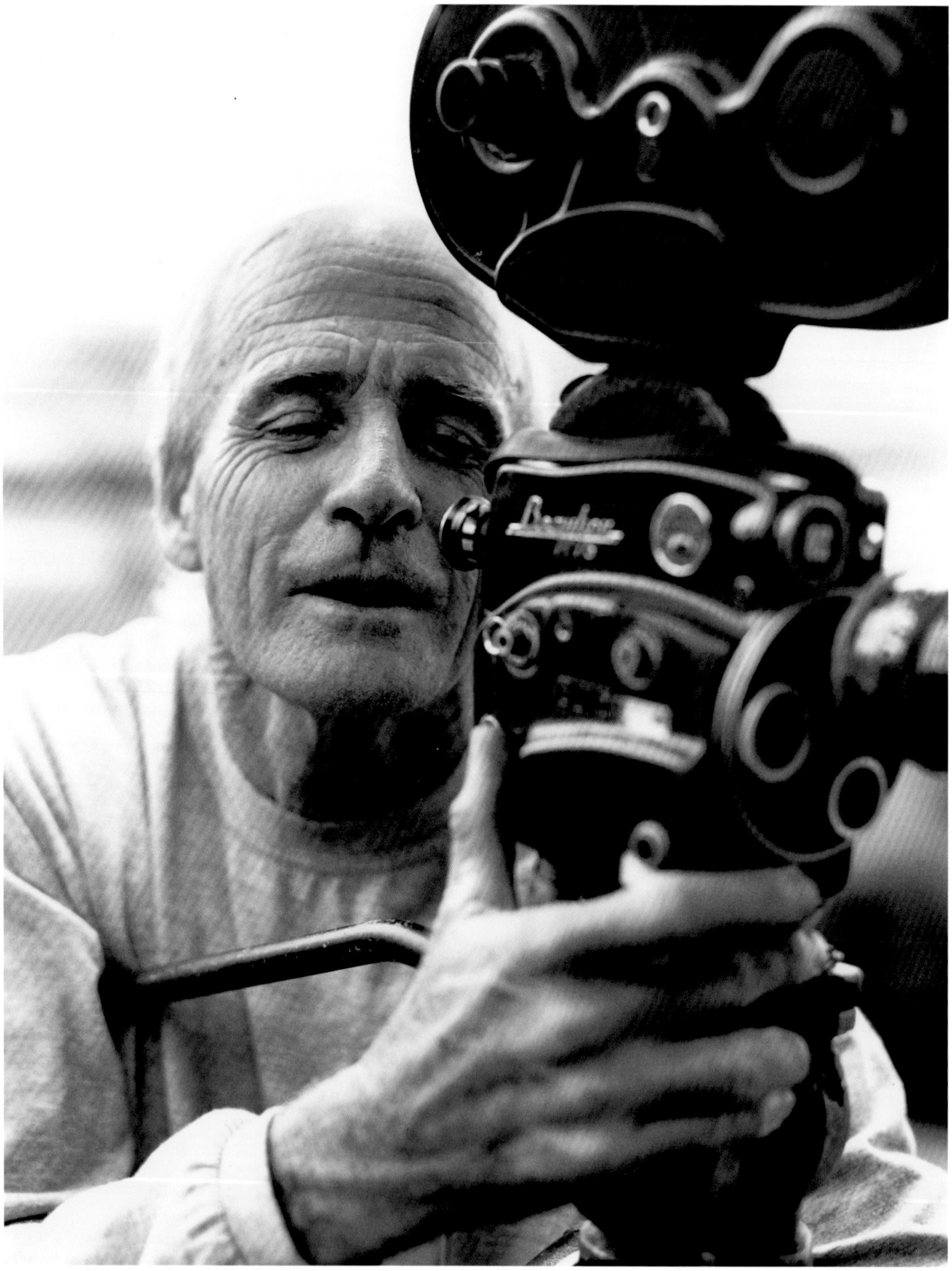

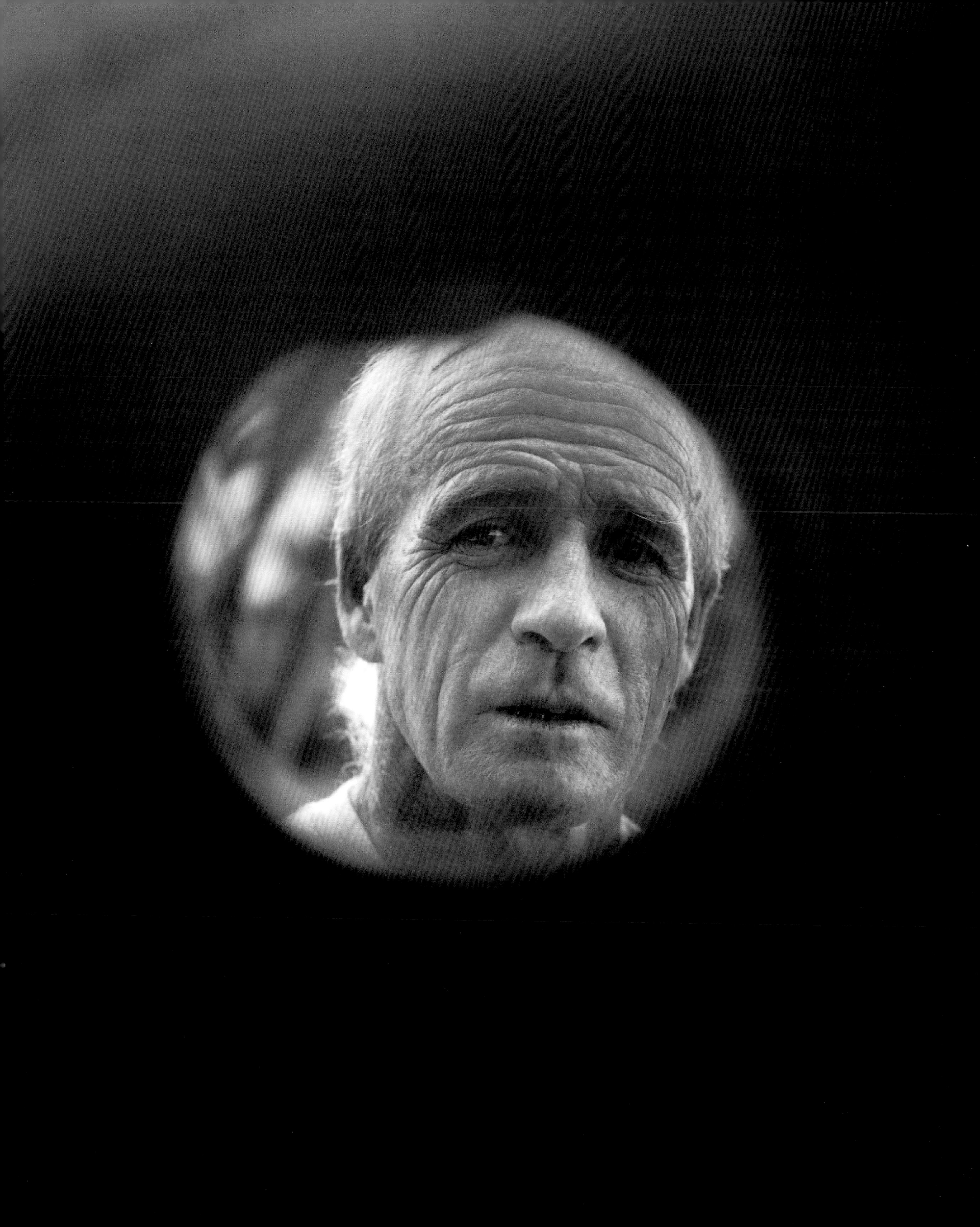

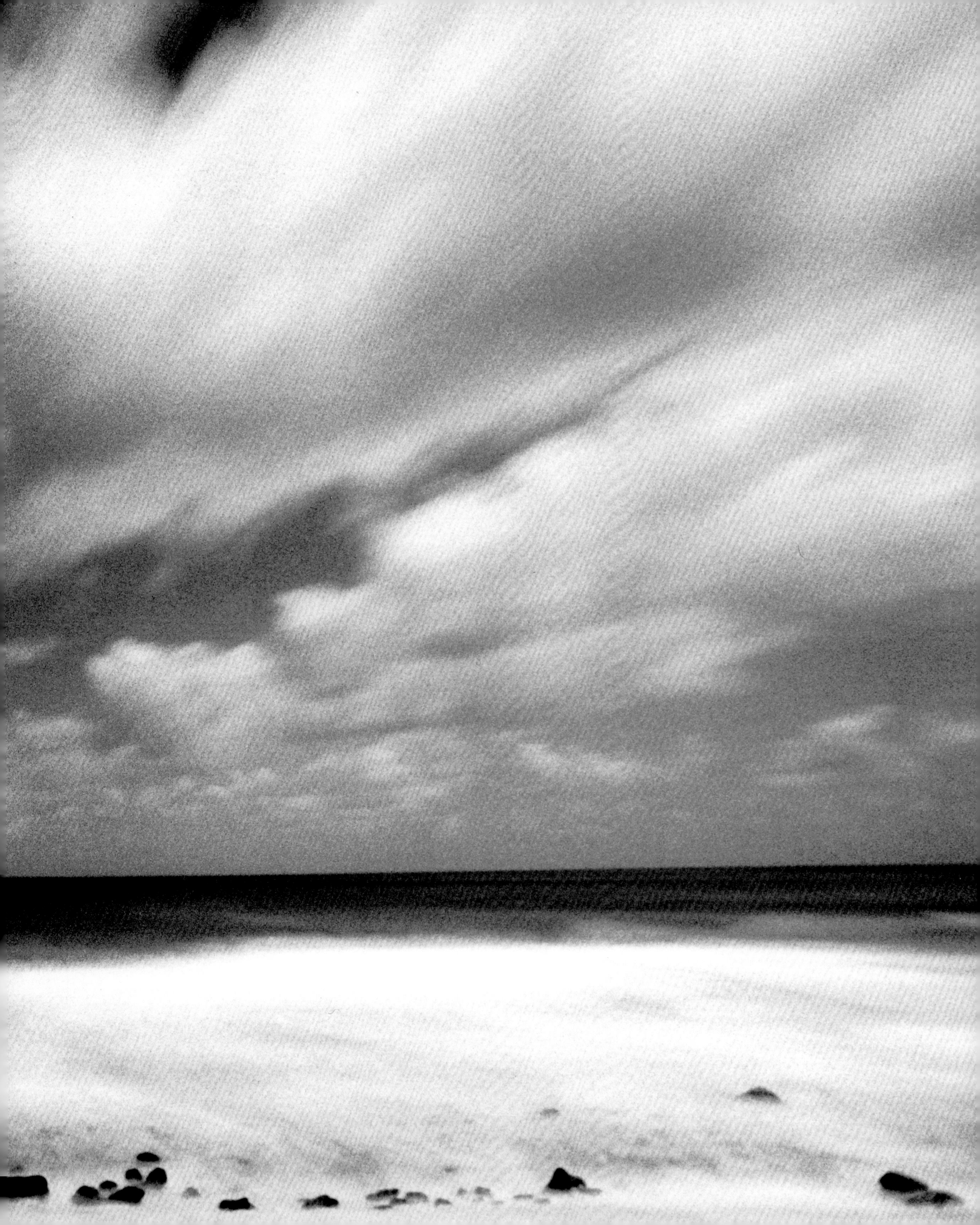

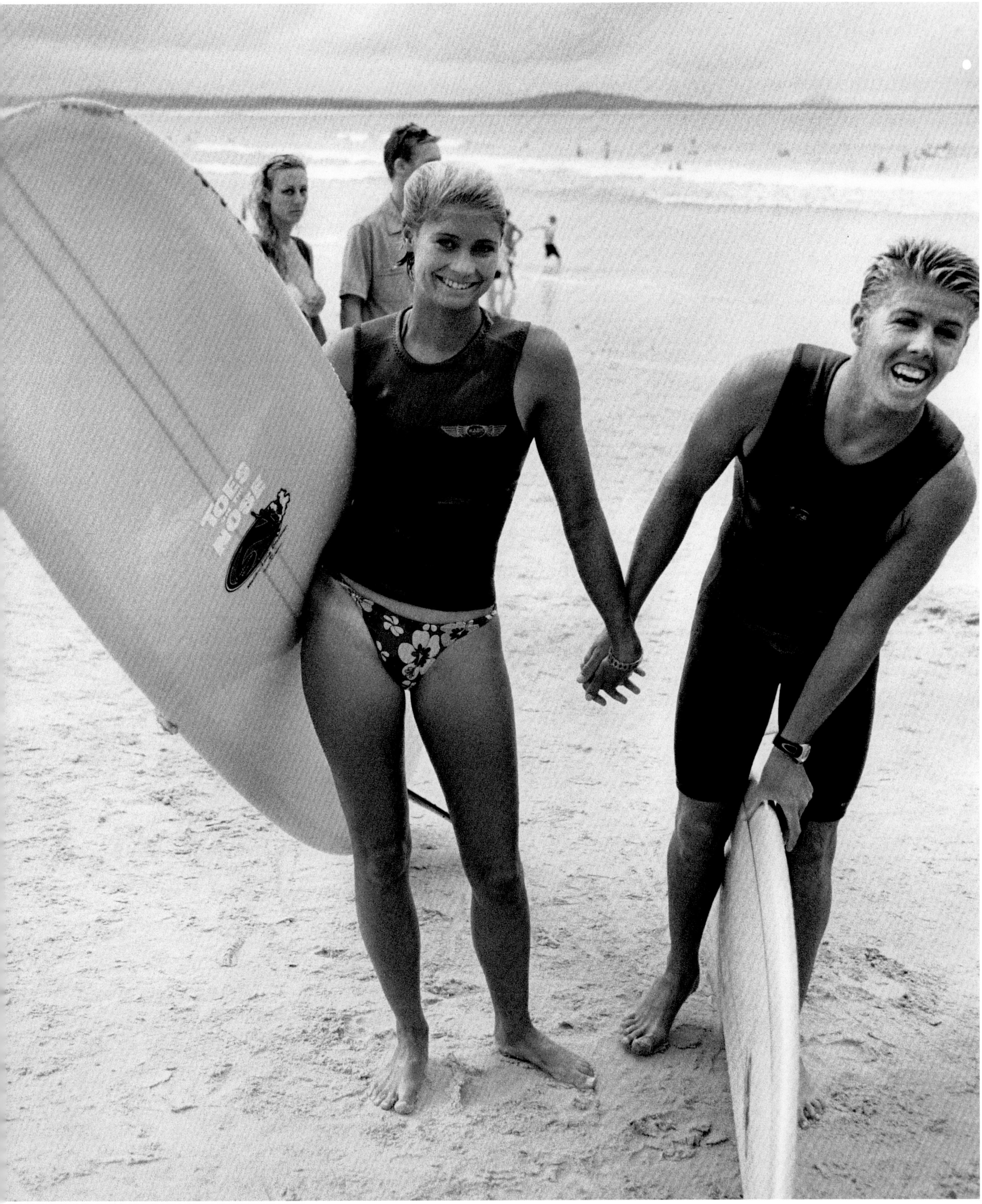
TOES
NOSE

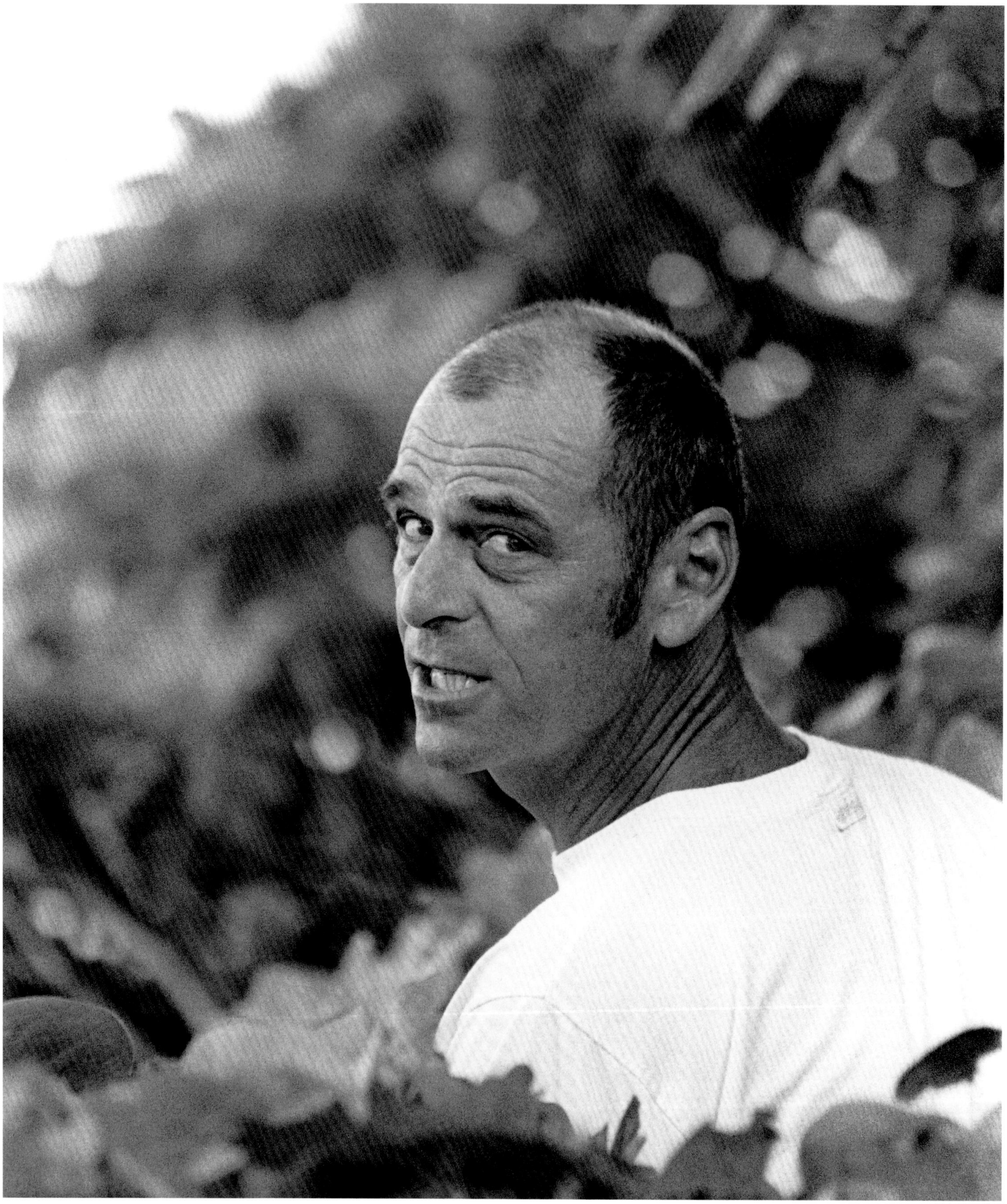

SKIP FRYE SURFBOARDS
CUSTOM DESIGN AND SHAPING

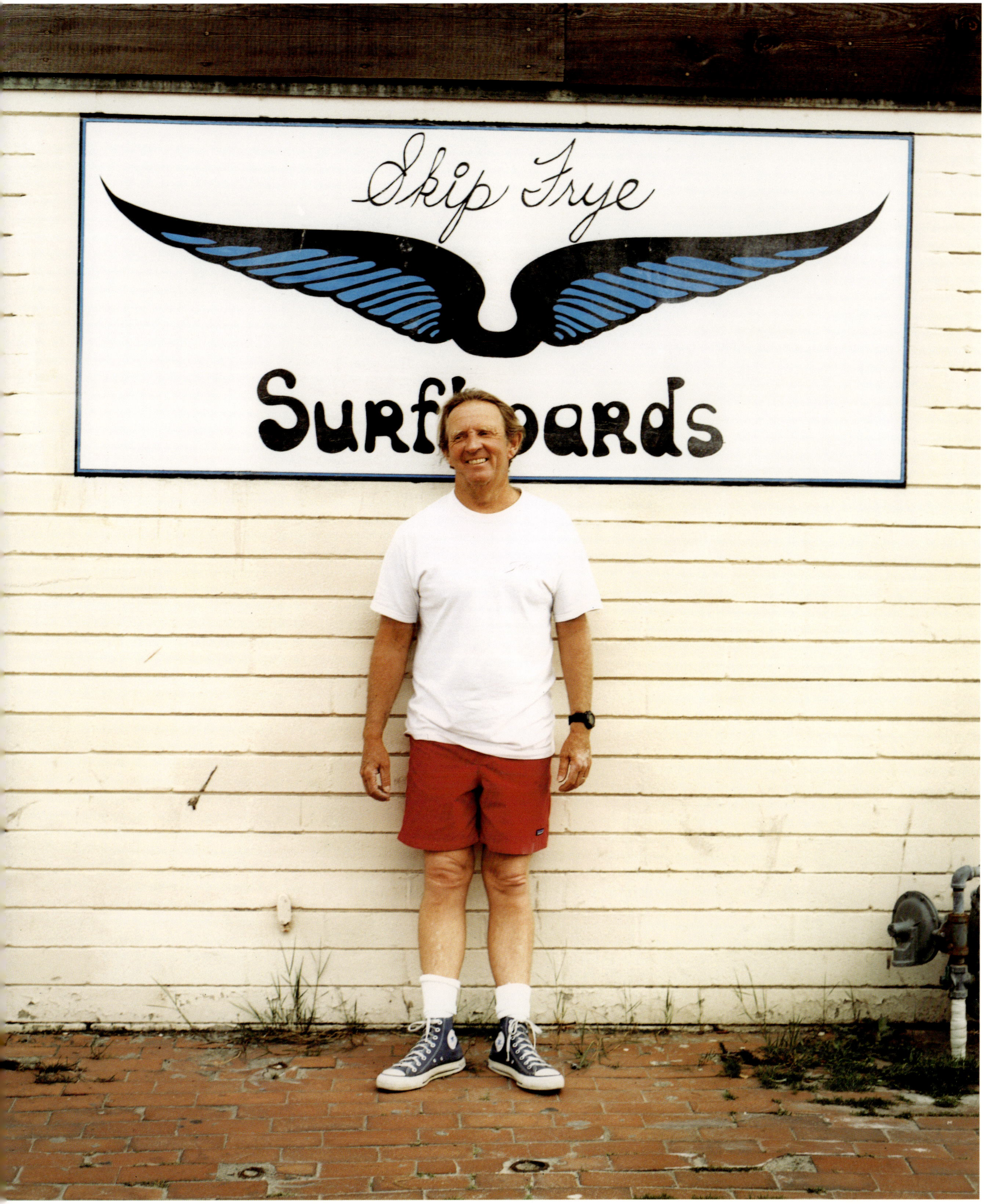
Skip Frye

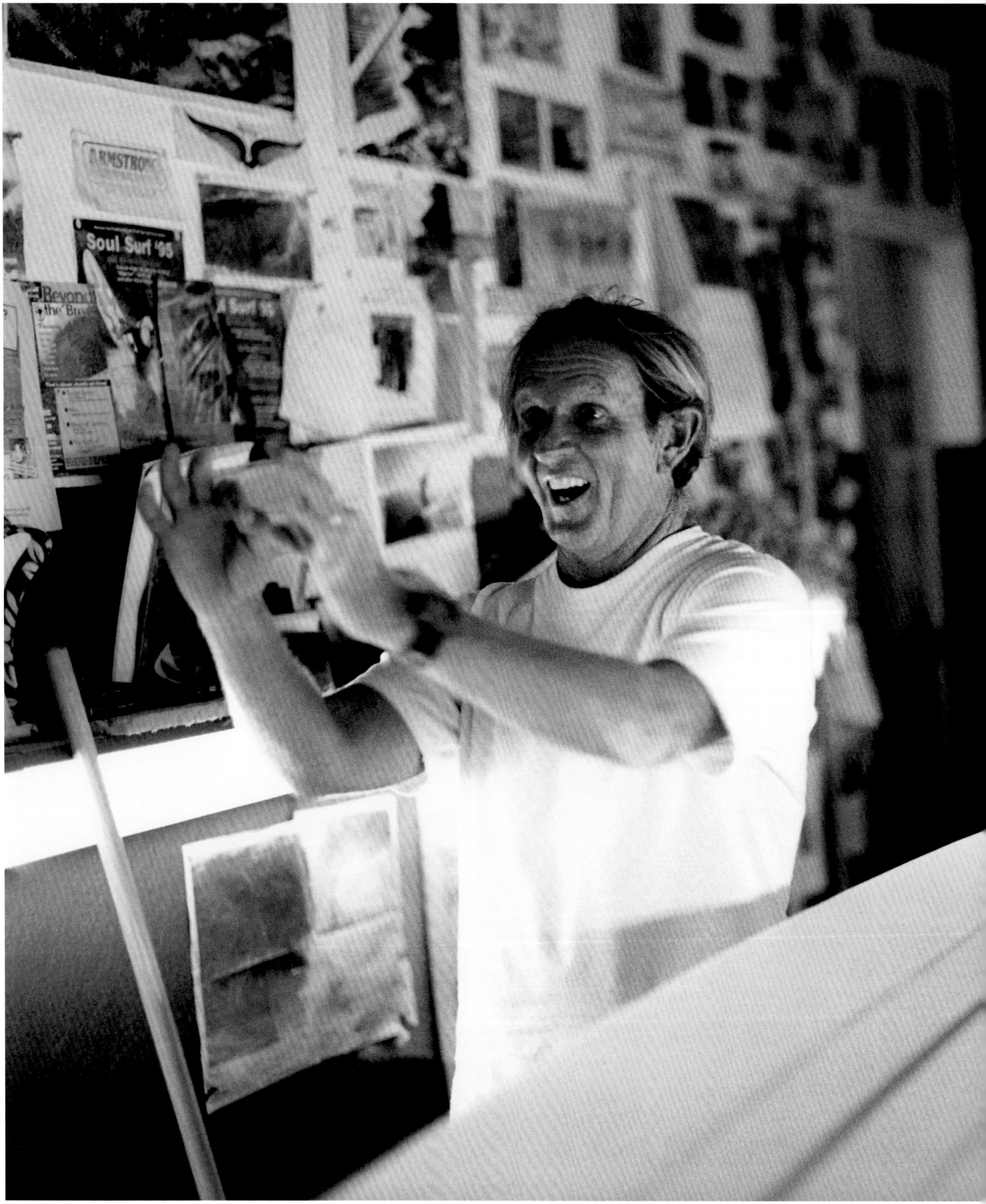
Soul Surf '95

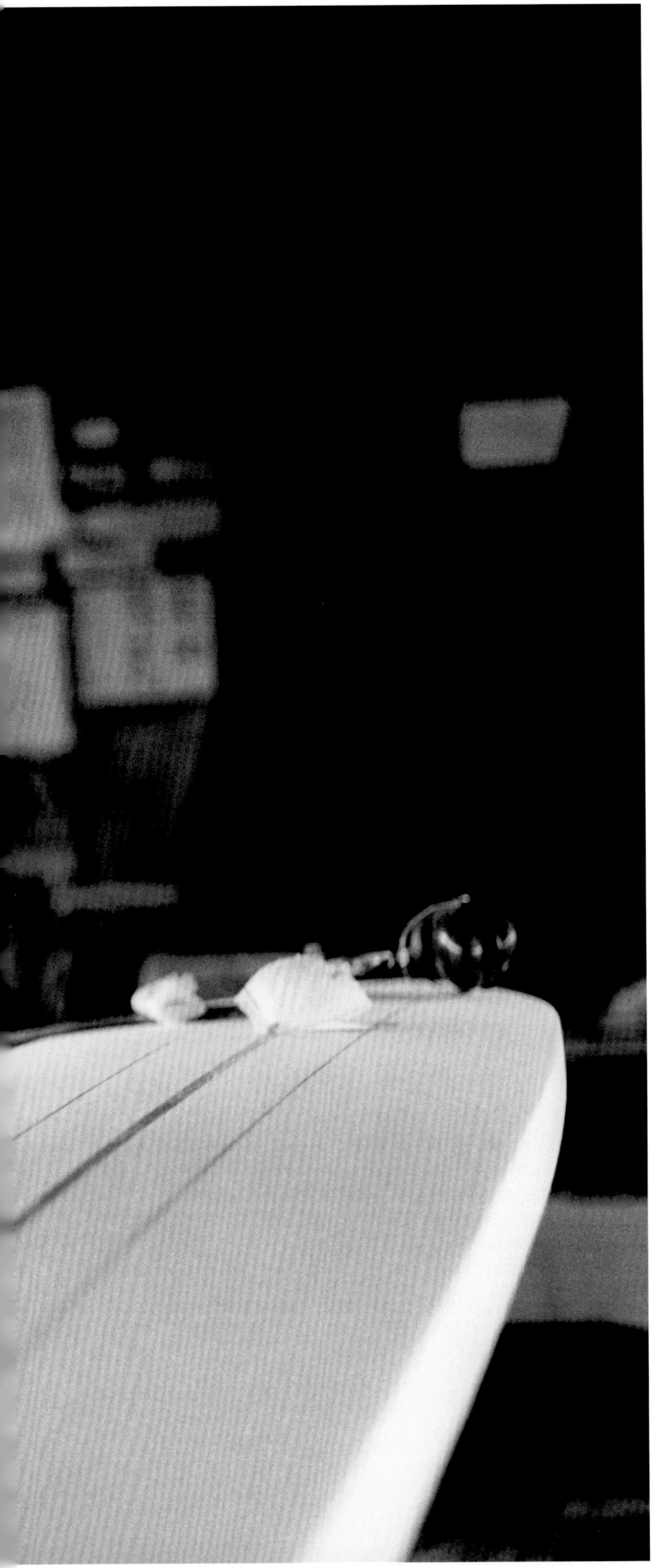

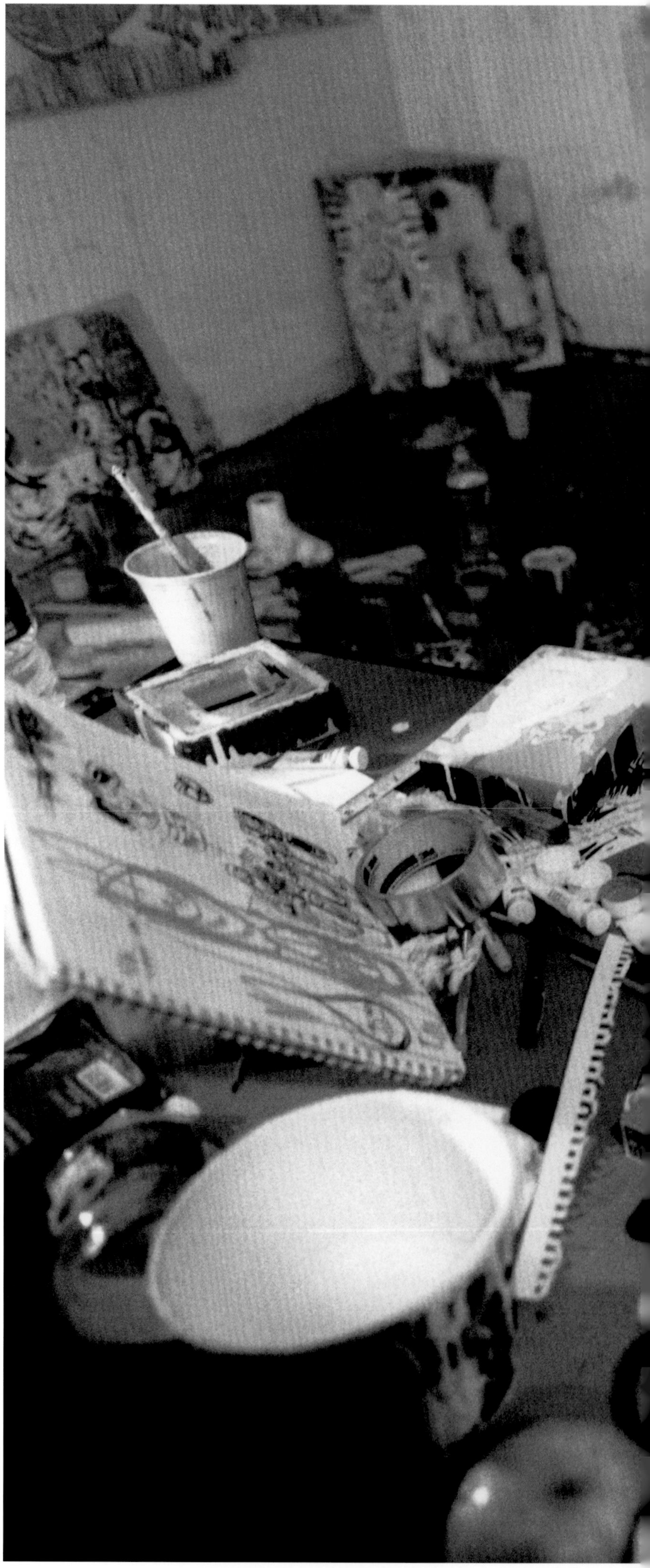

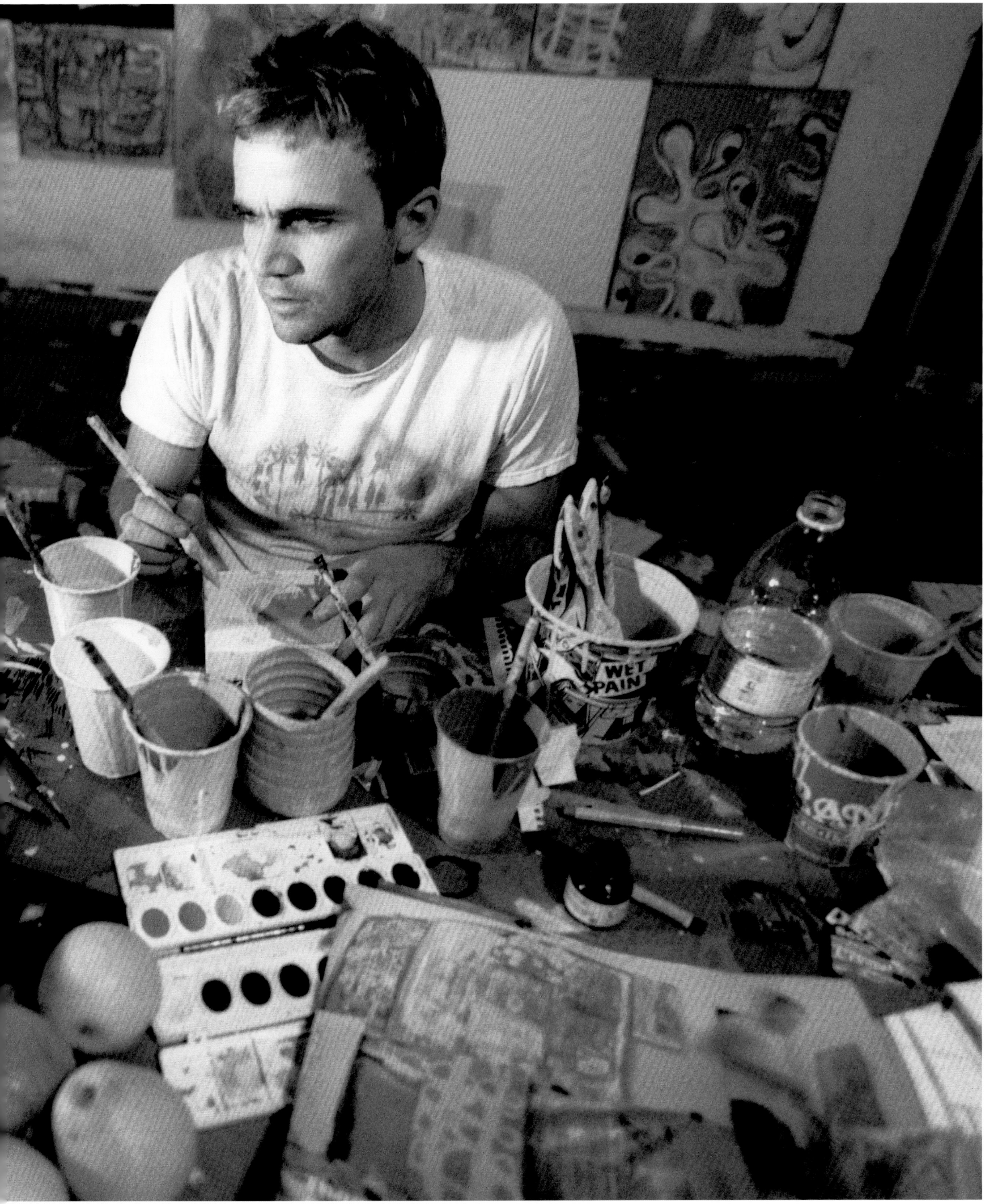
WET PAINT

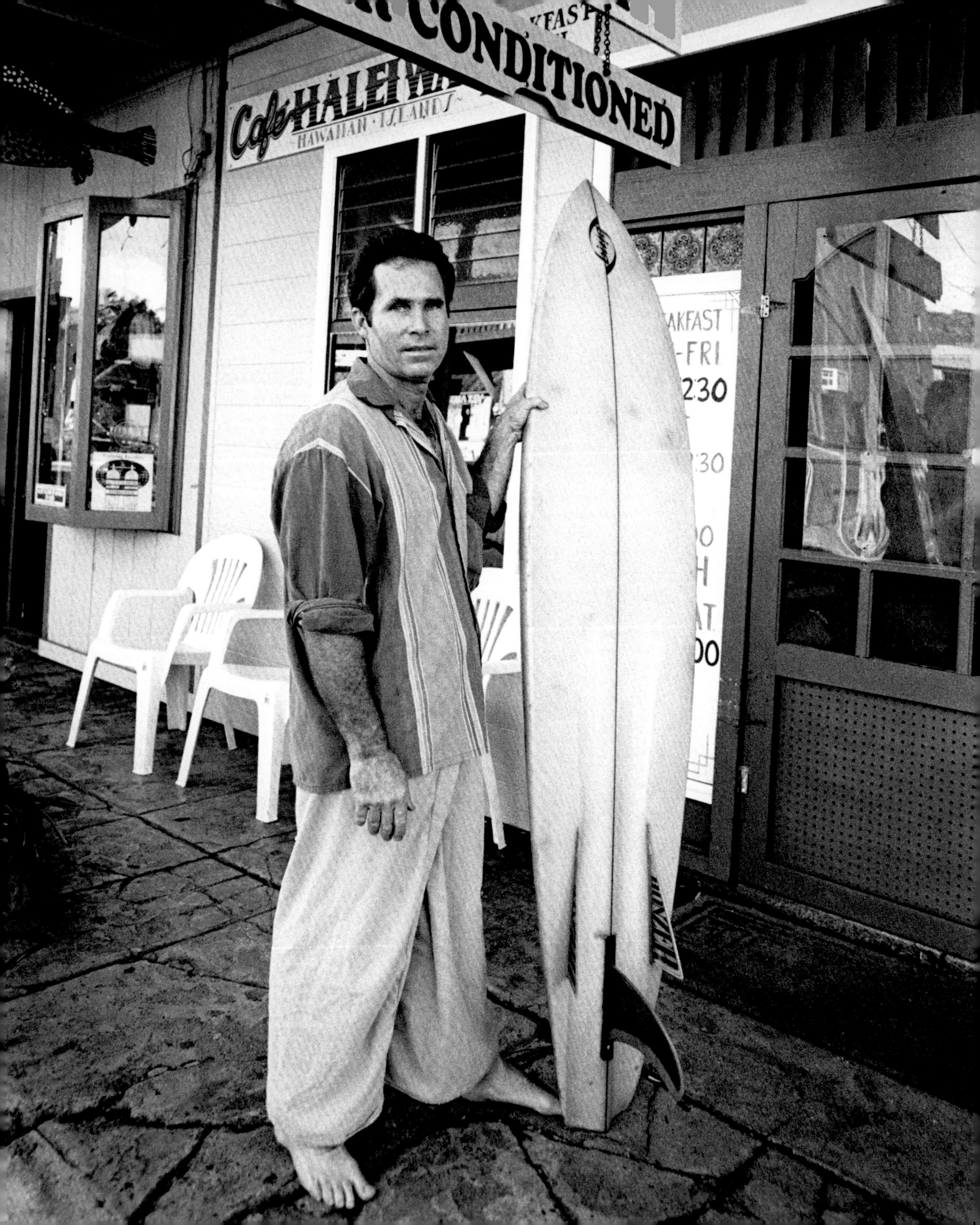
CONDITIONED
Cafe
HAWAIIAN · ISLANDS

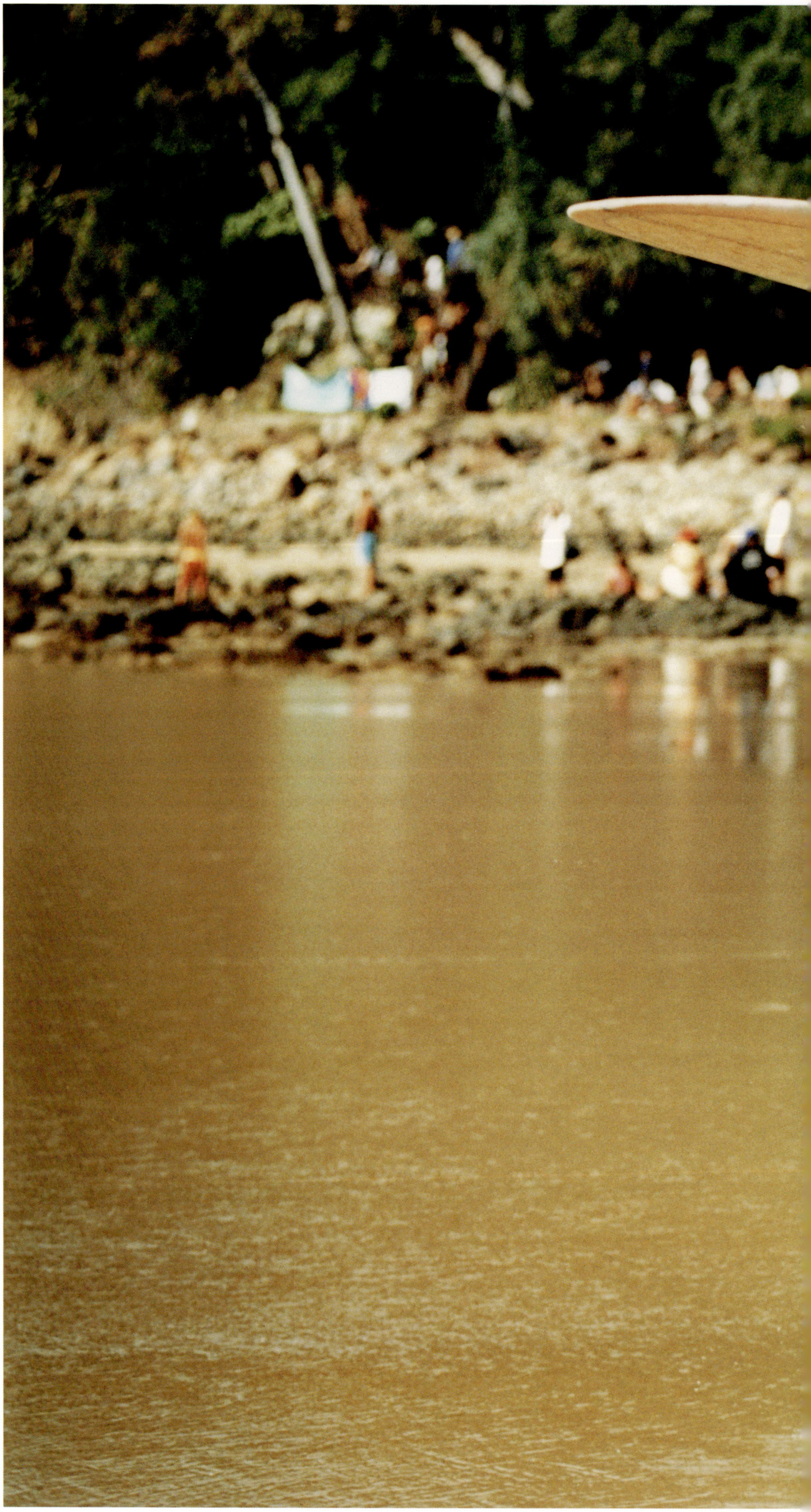

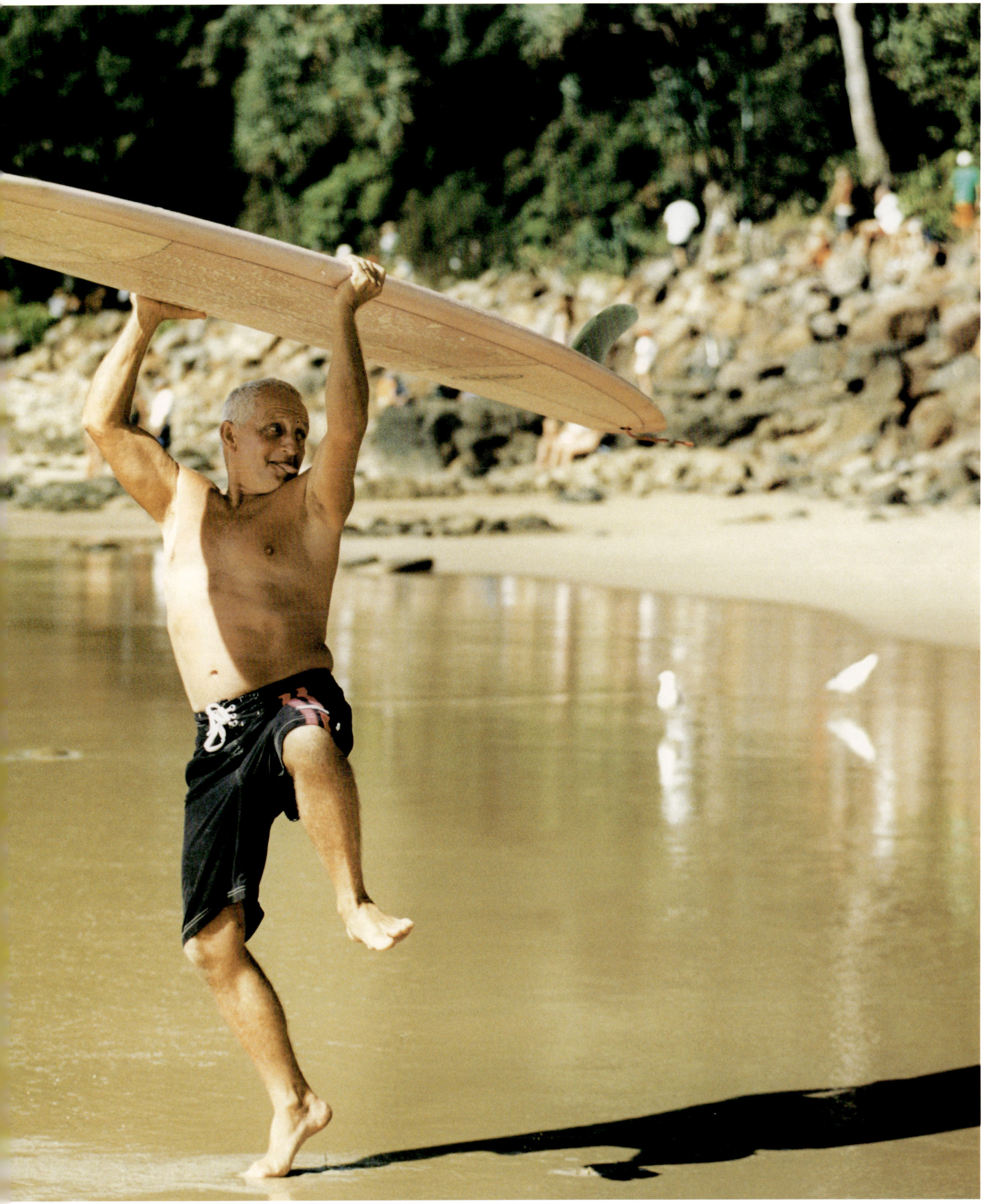

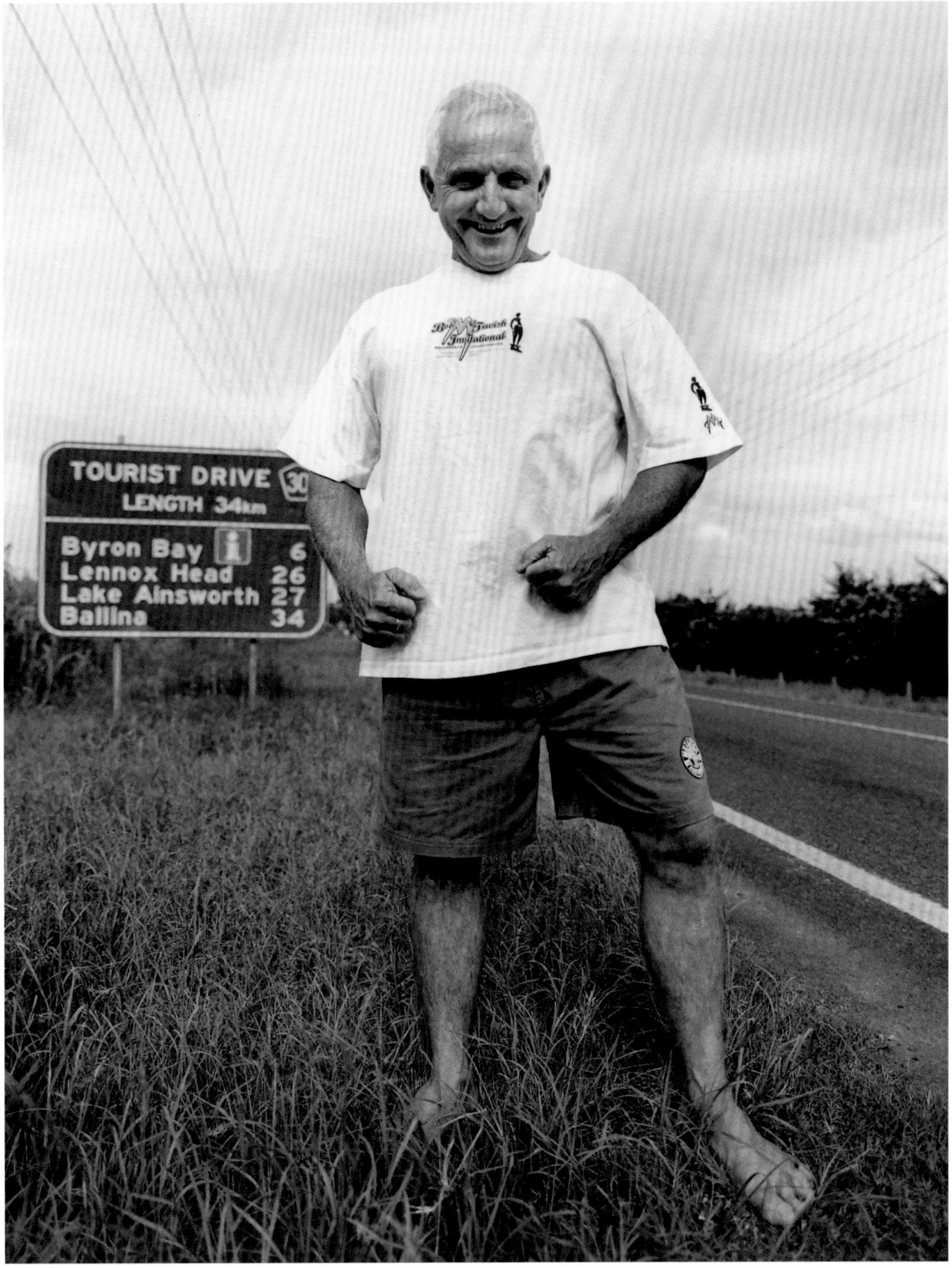
TOURIST DRIVE 30
LENGTH 34km
Byron Bay 6
Lennox Head 26
Lake Ainsworth 27
Ballina 34

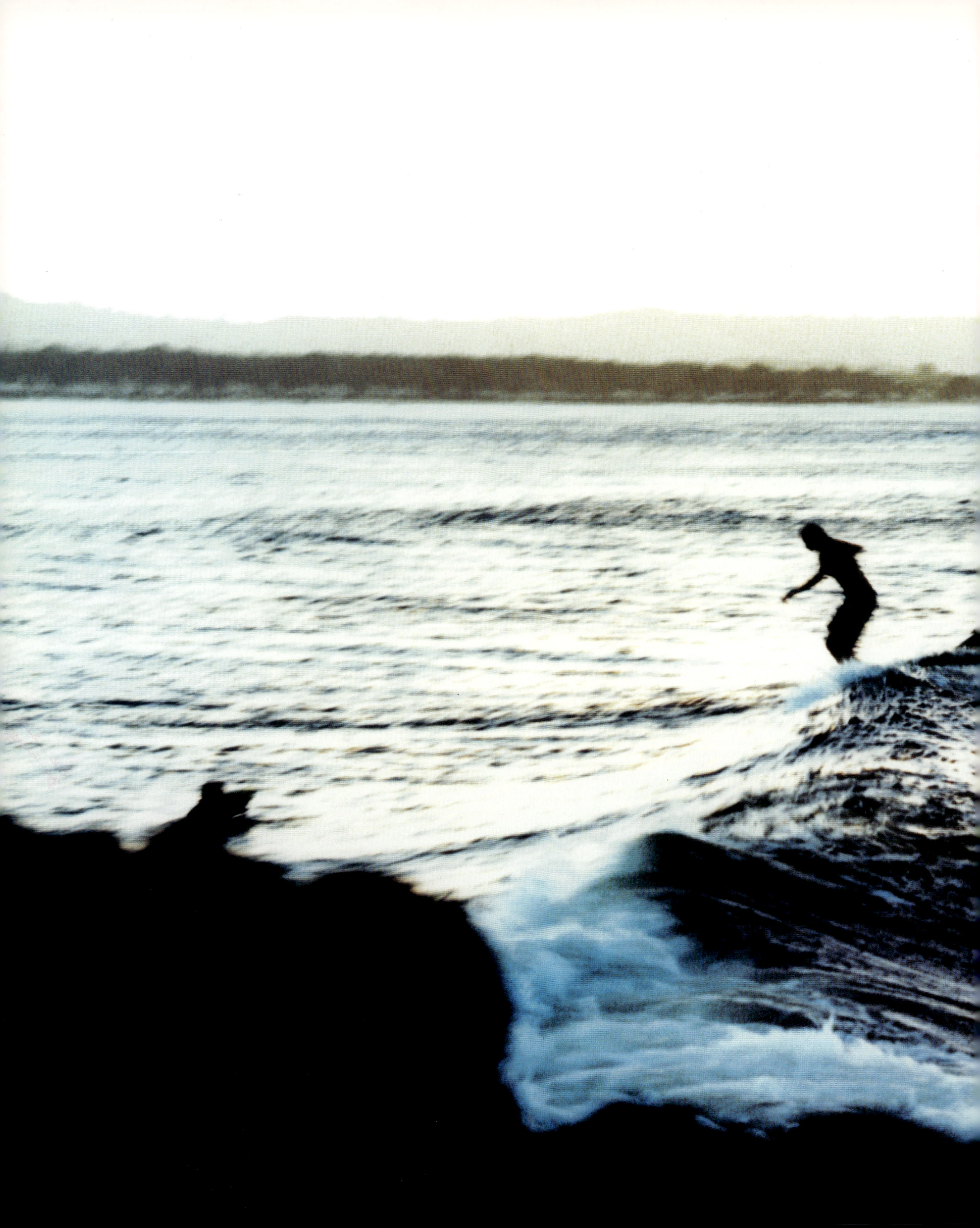

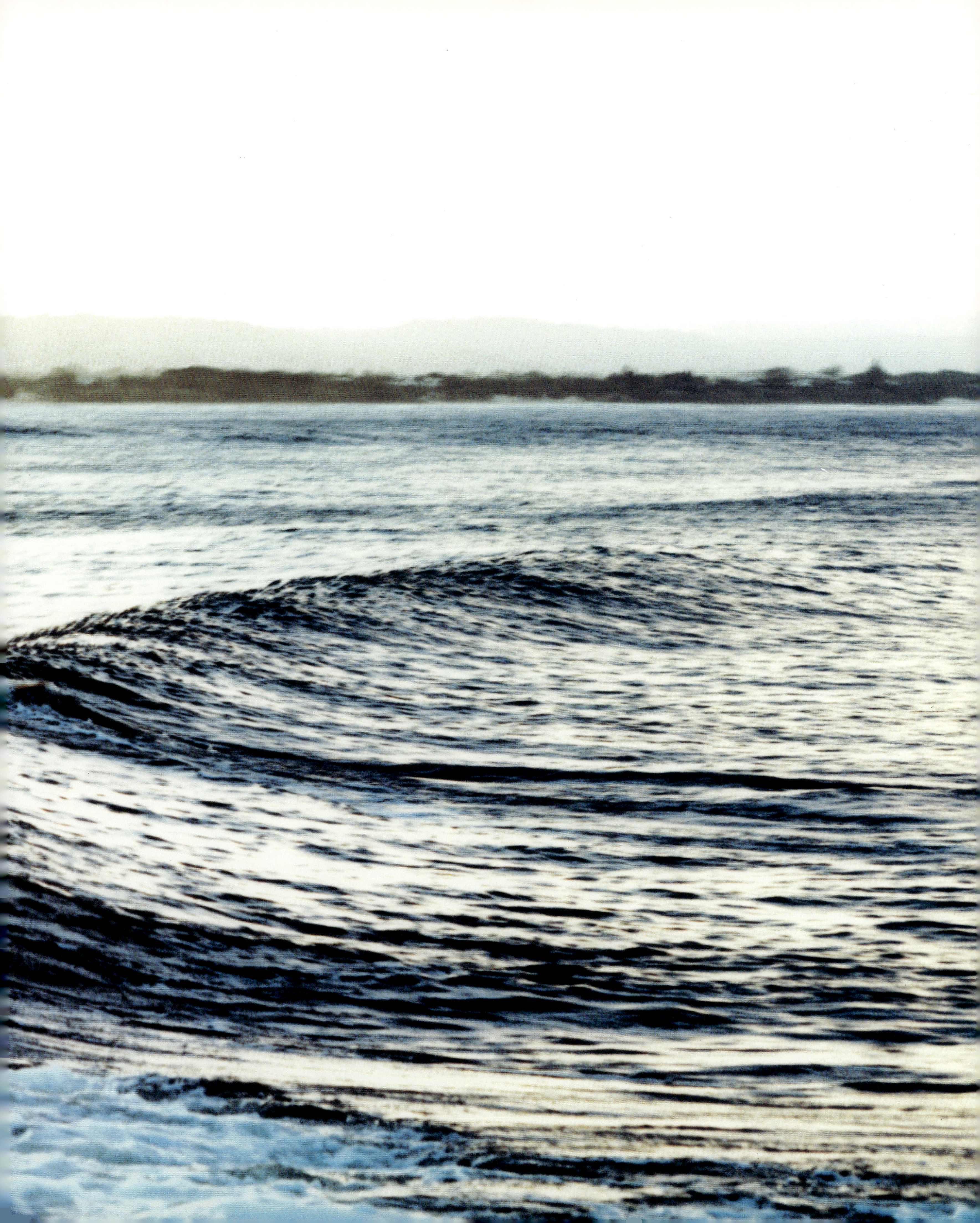

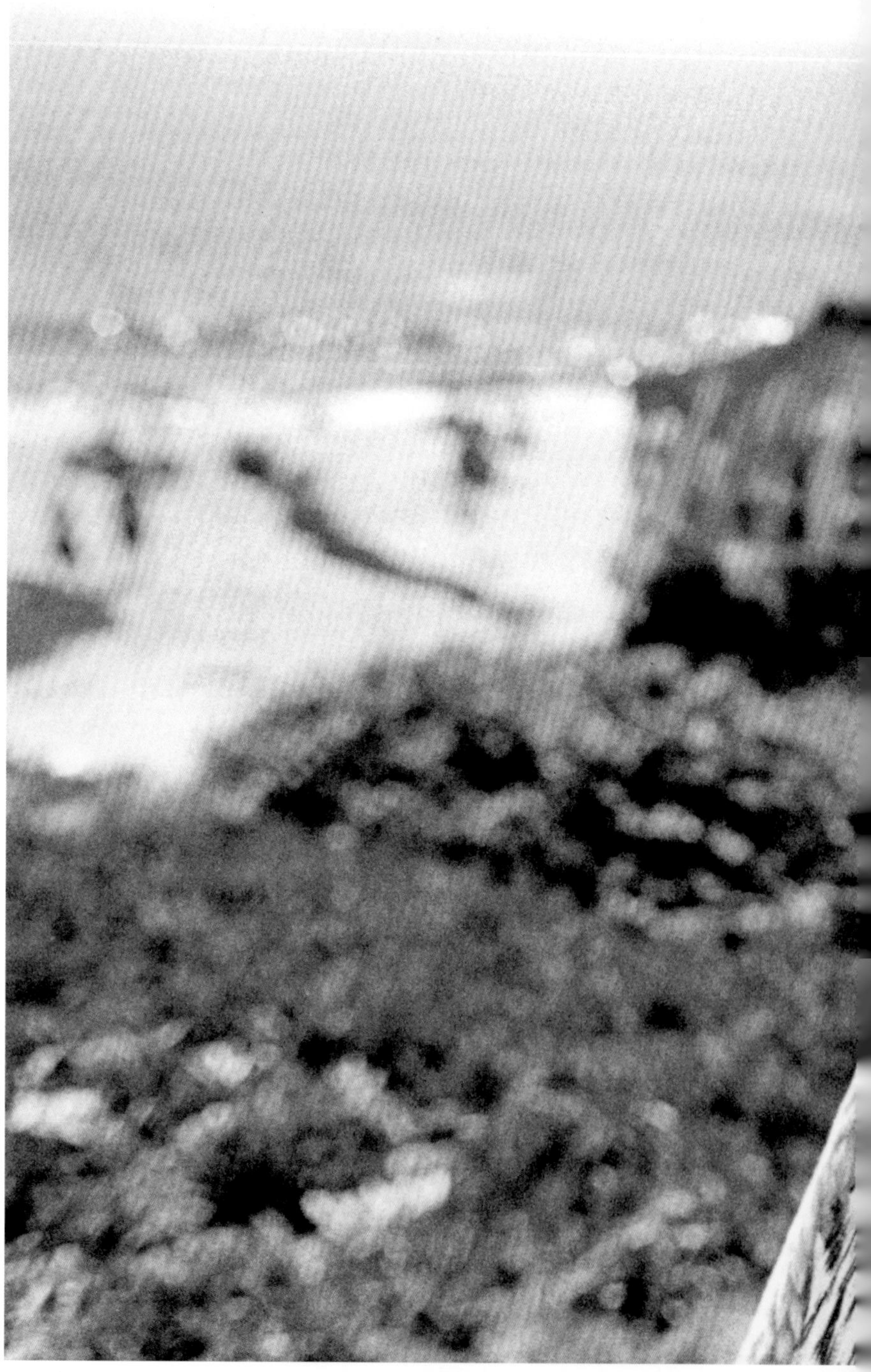

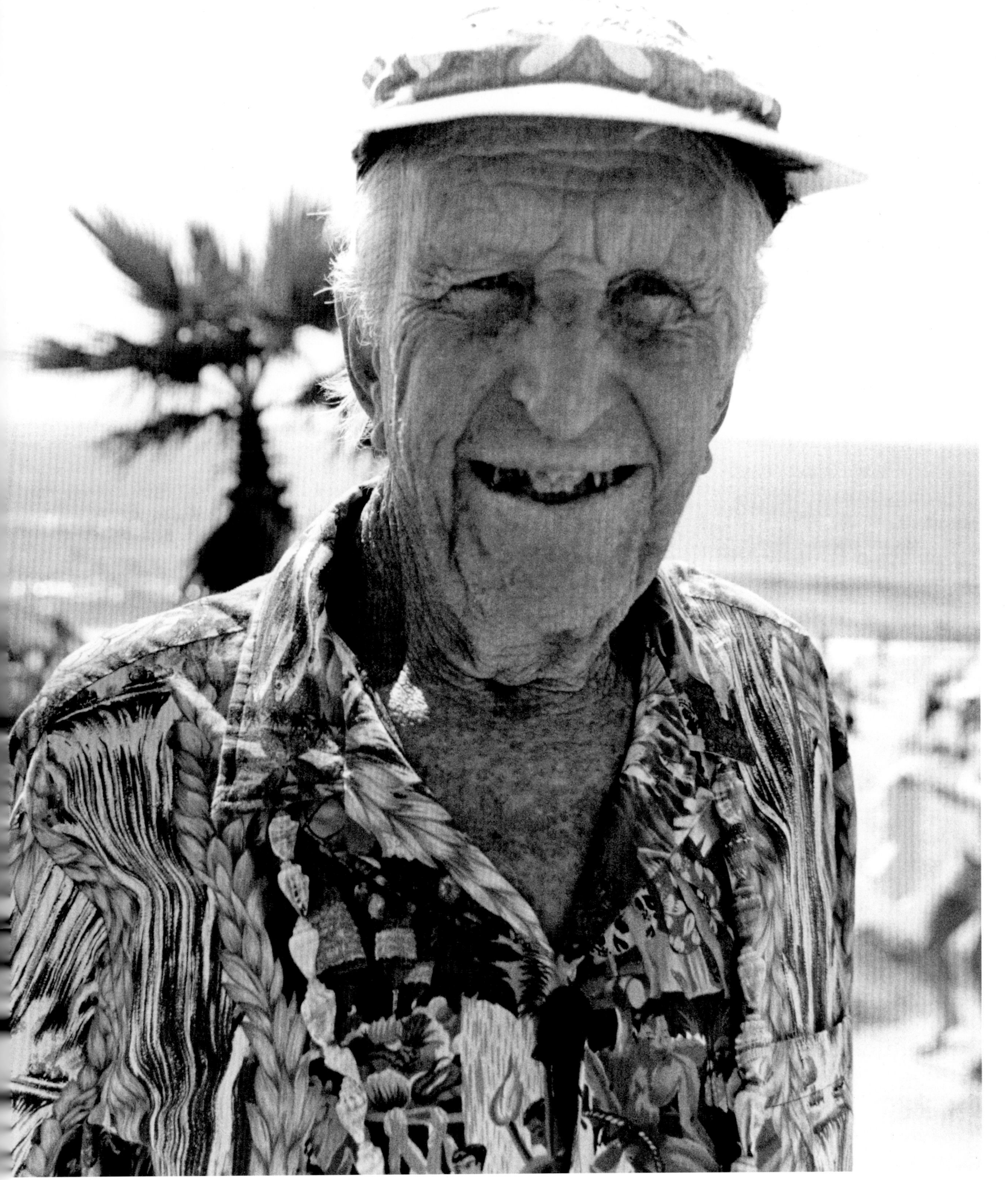

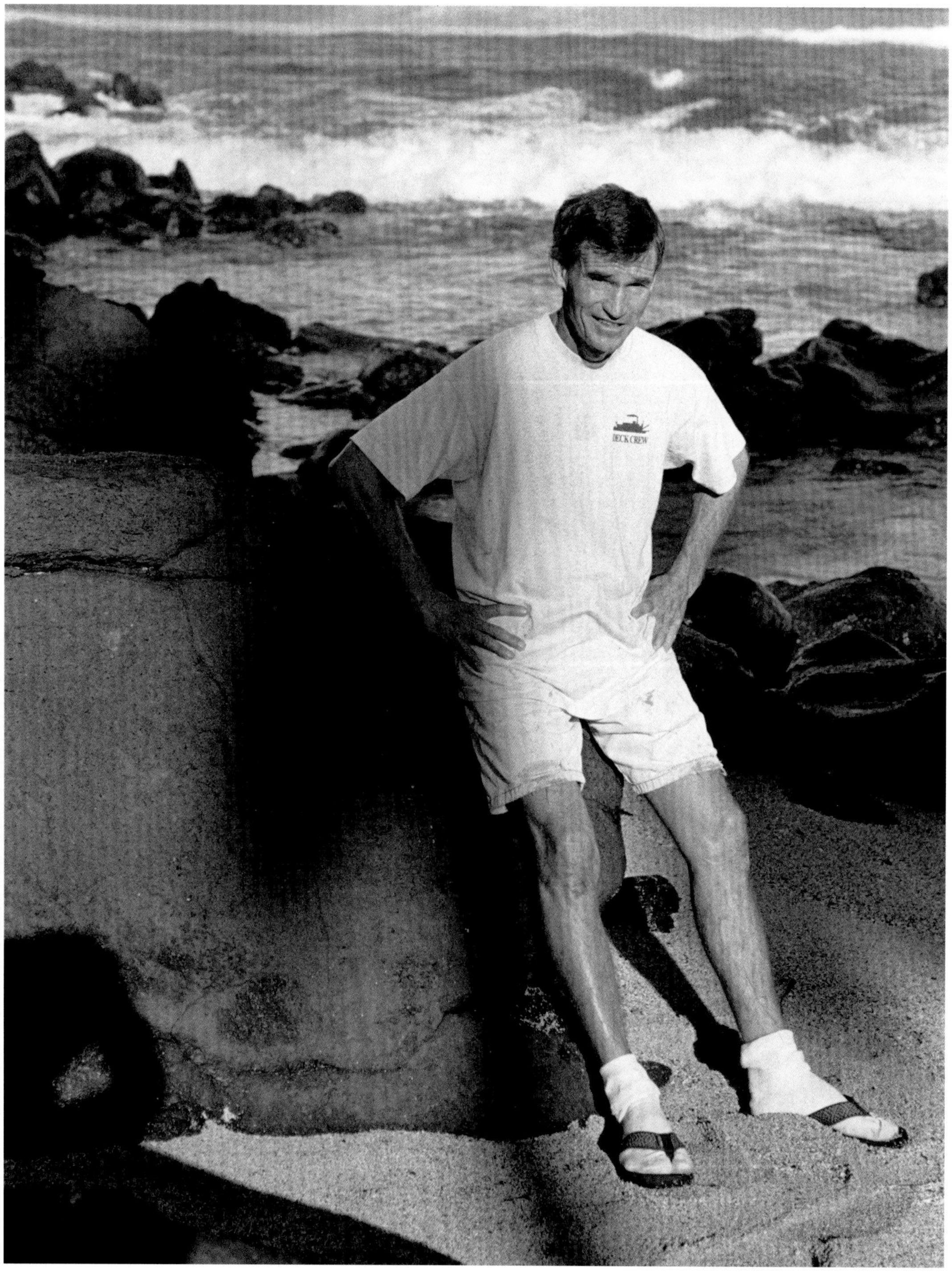
DECK CREW

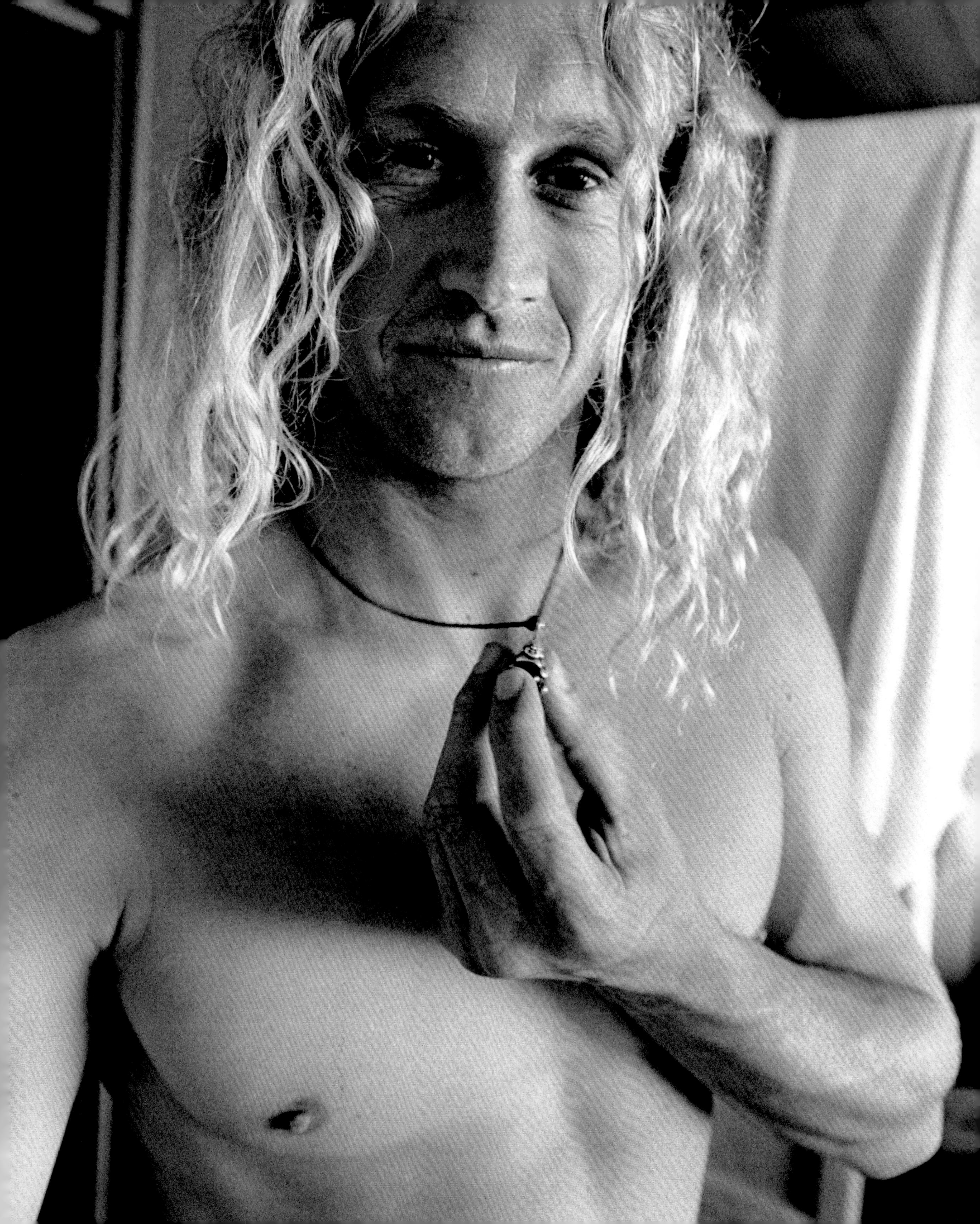

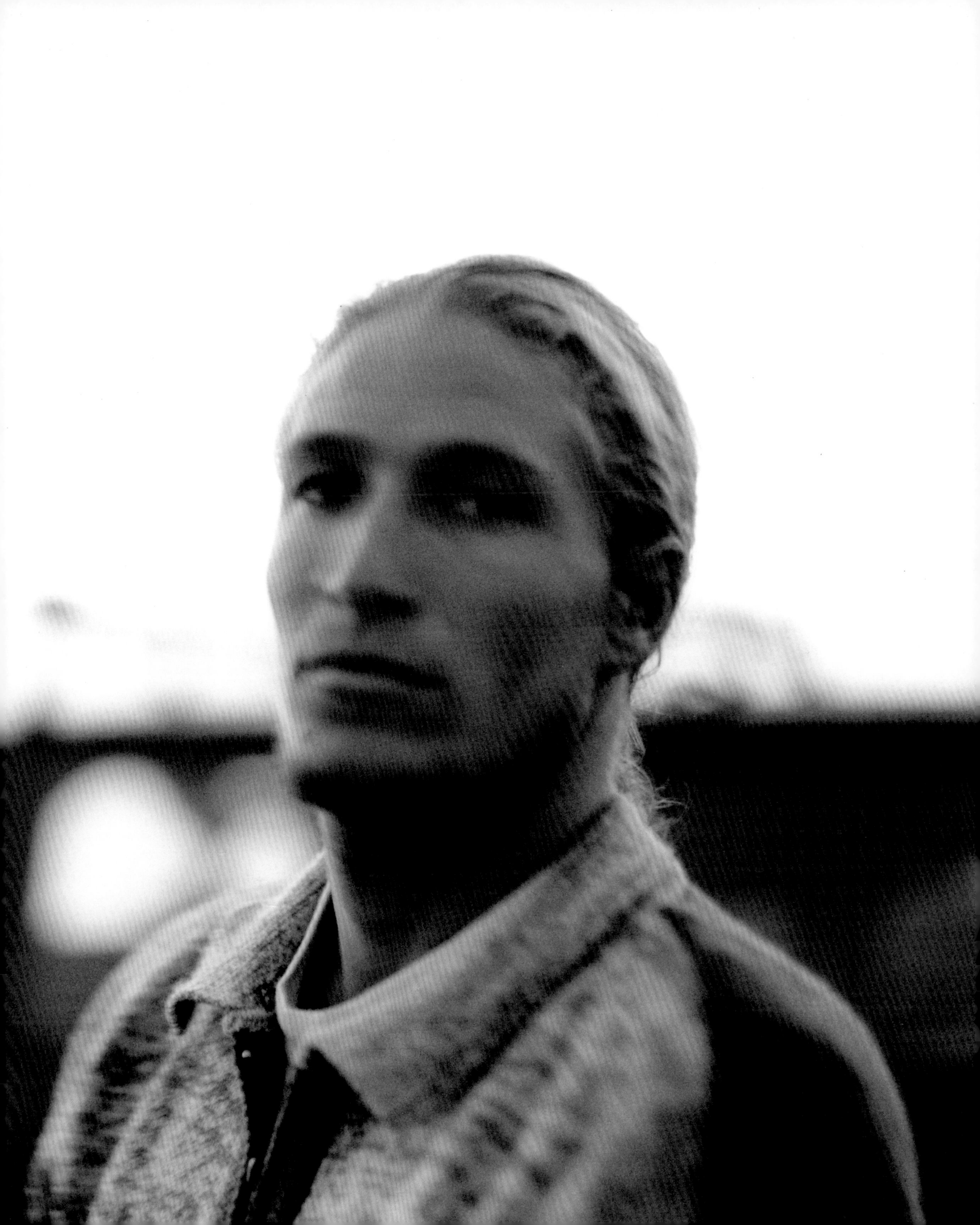

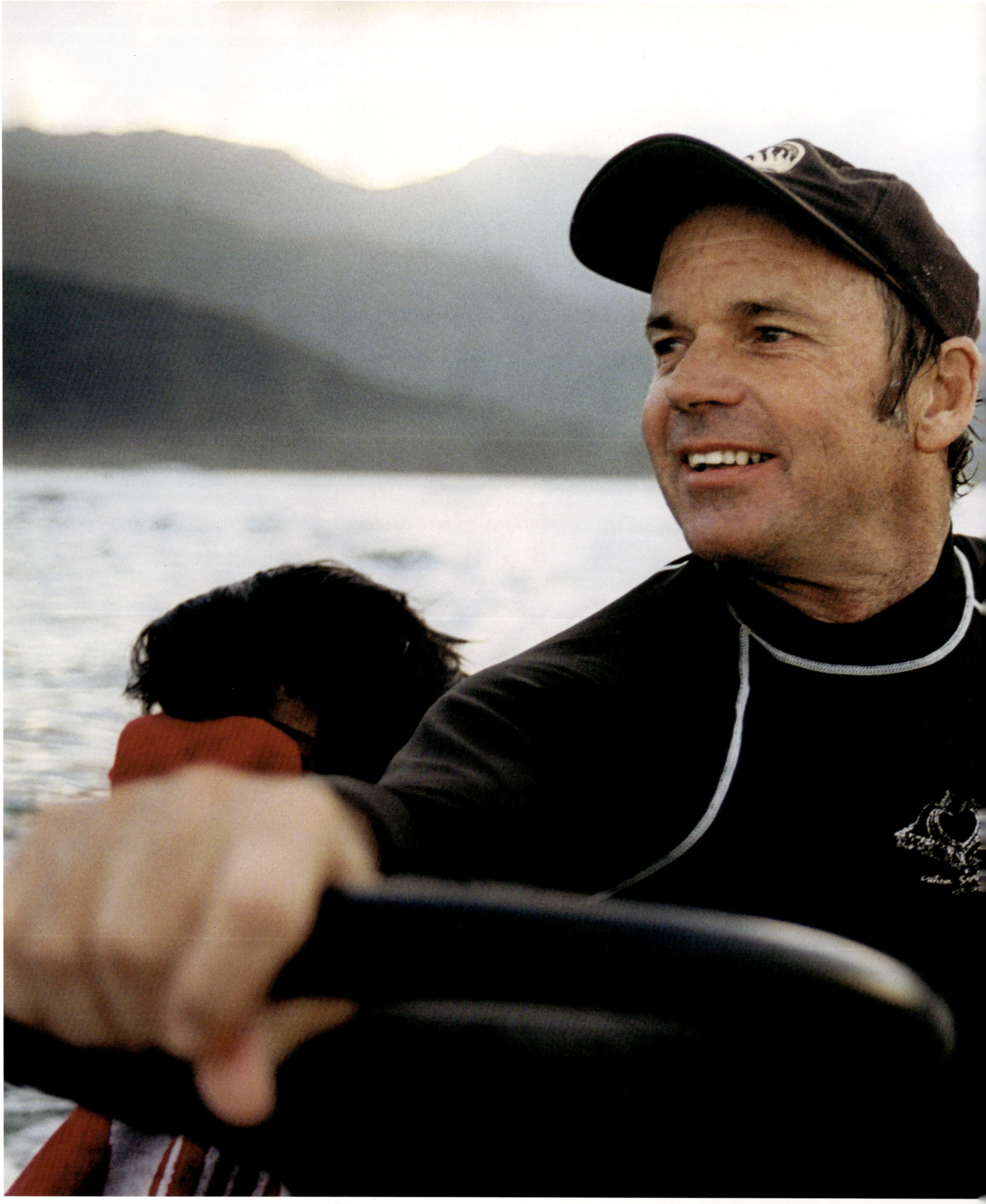

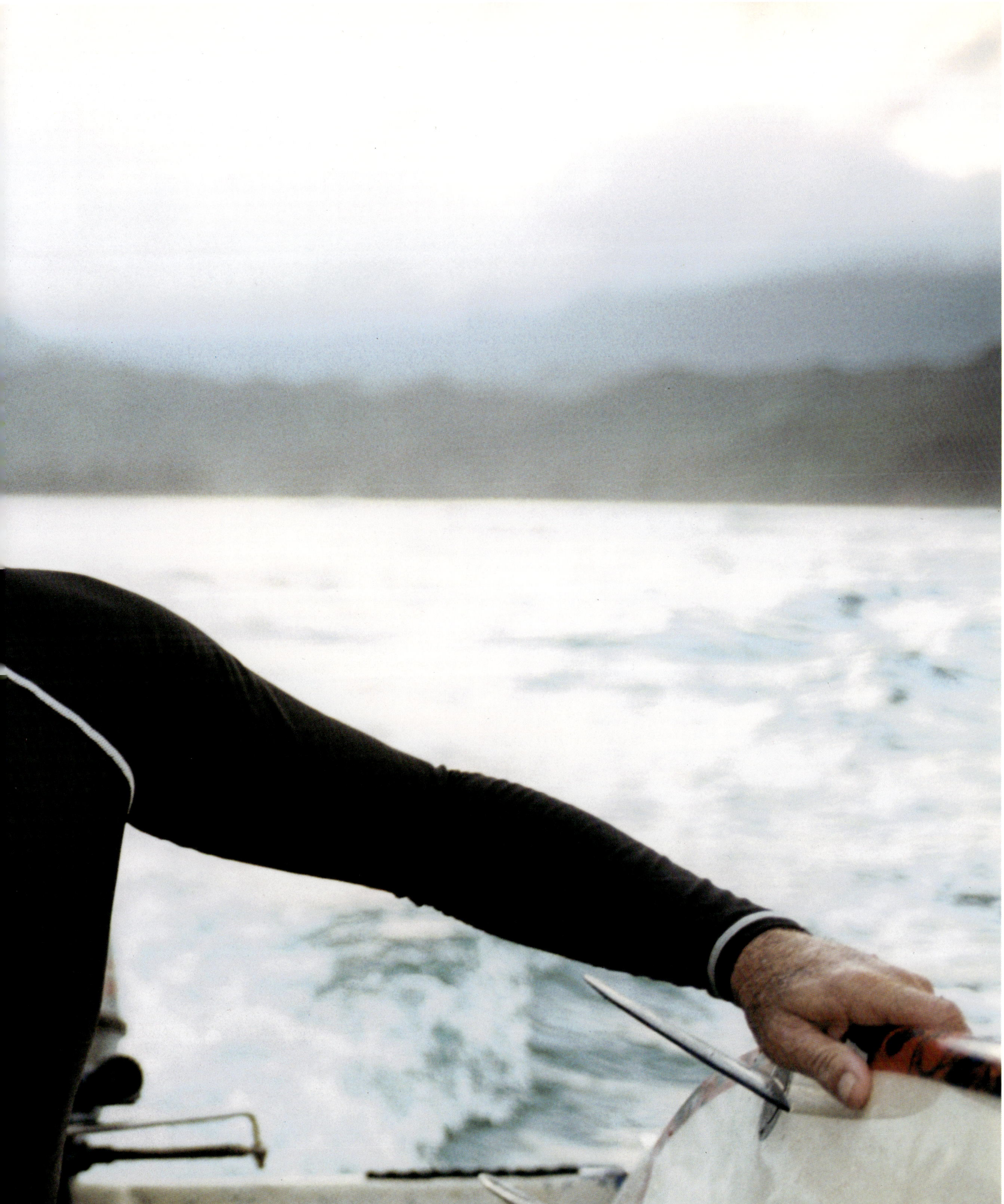

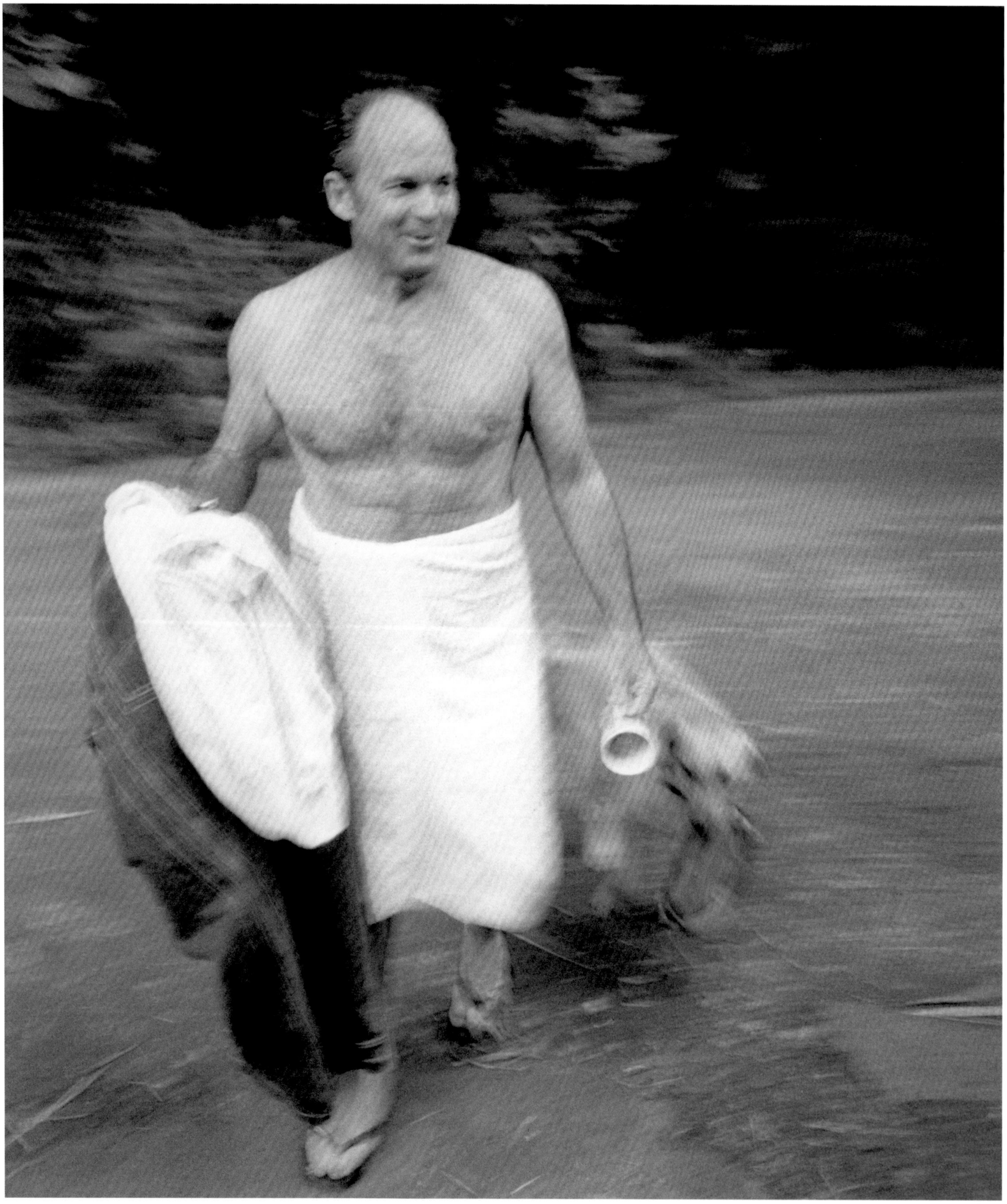

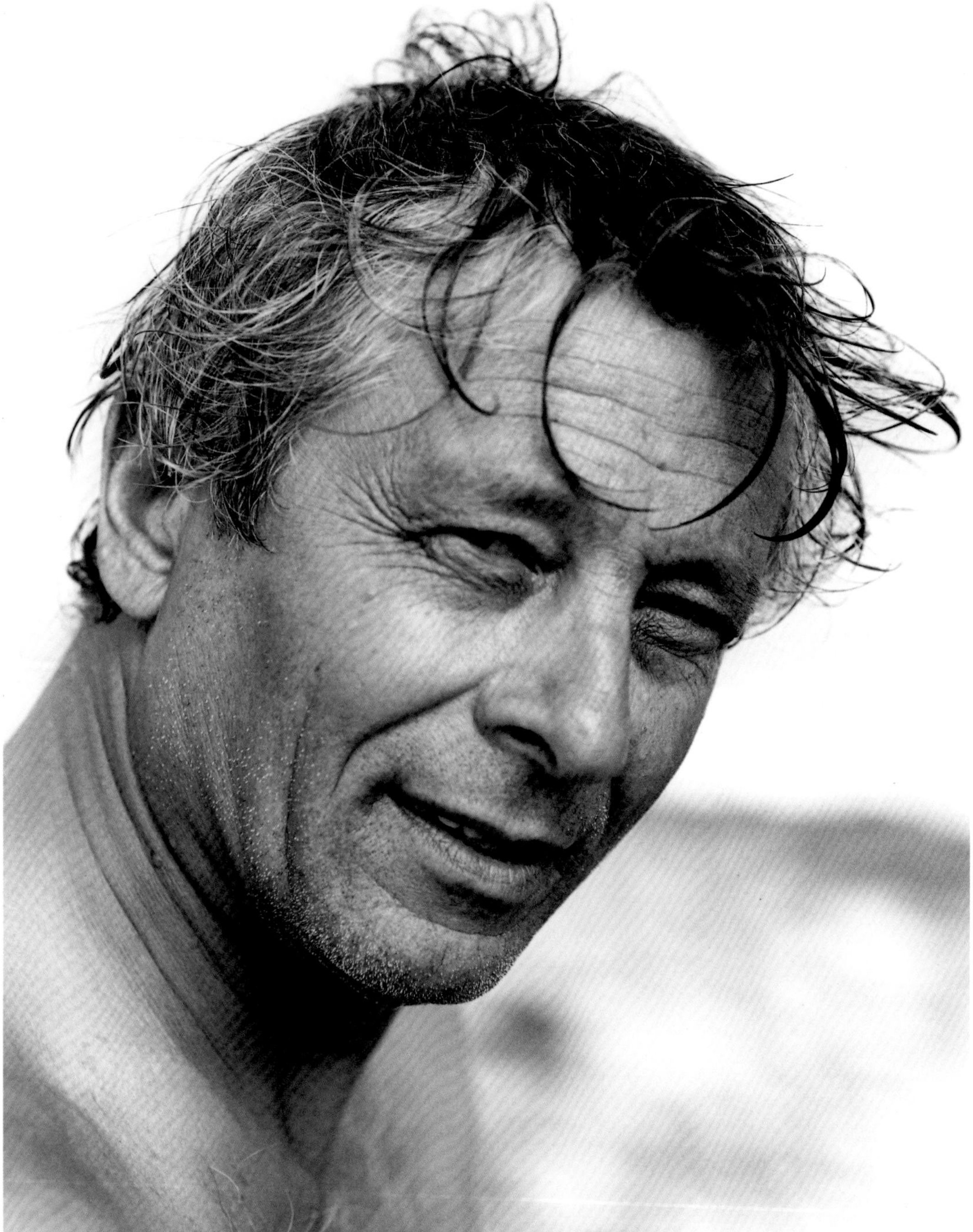

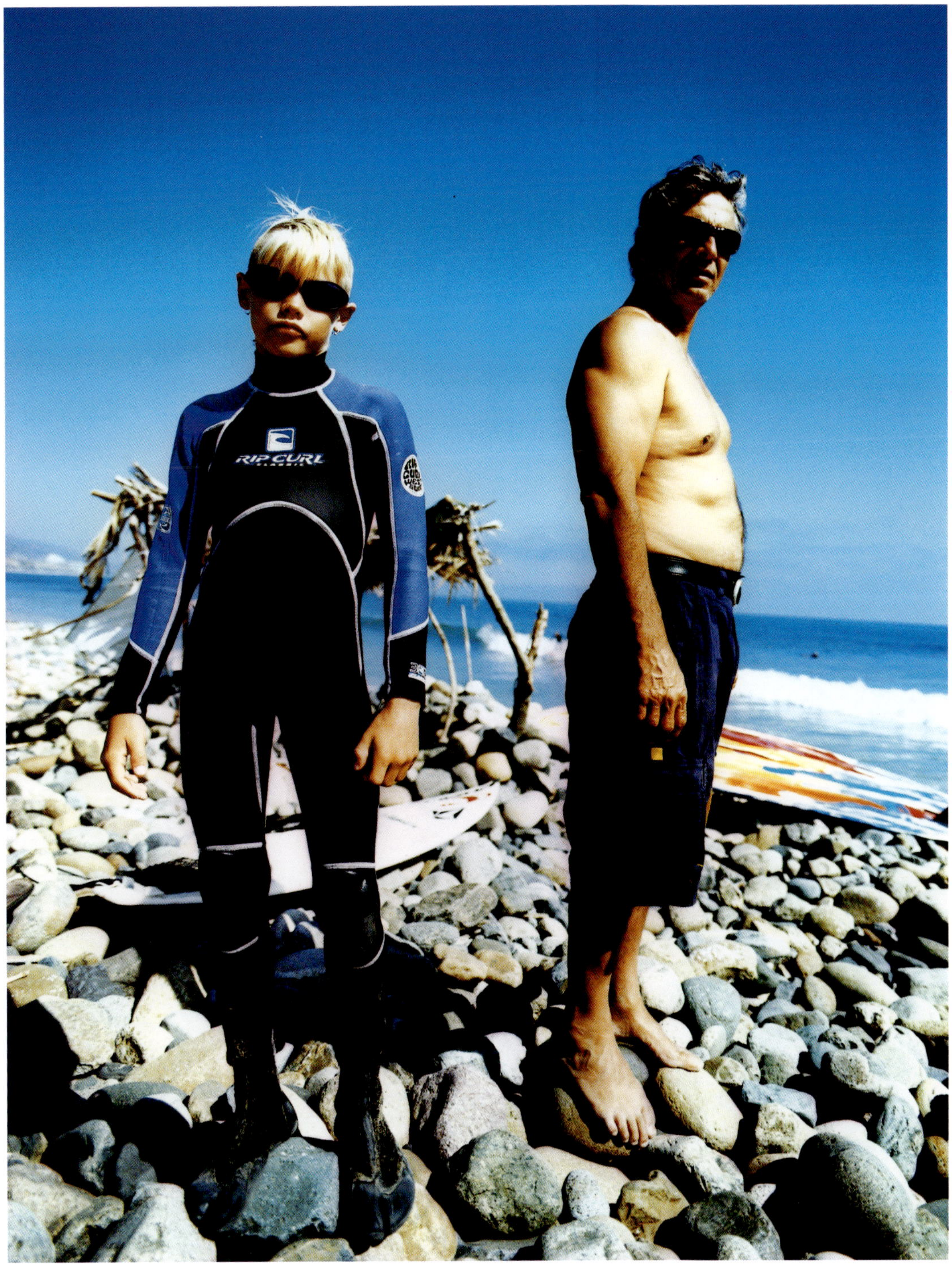
RIP CURL

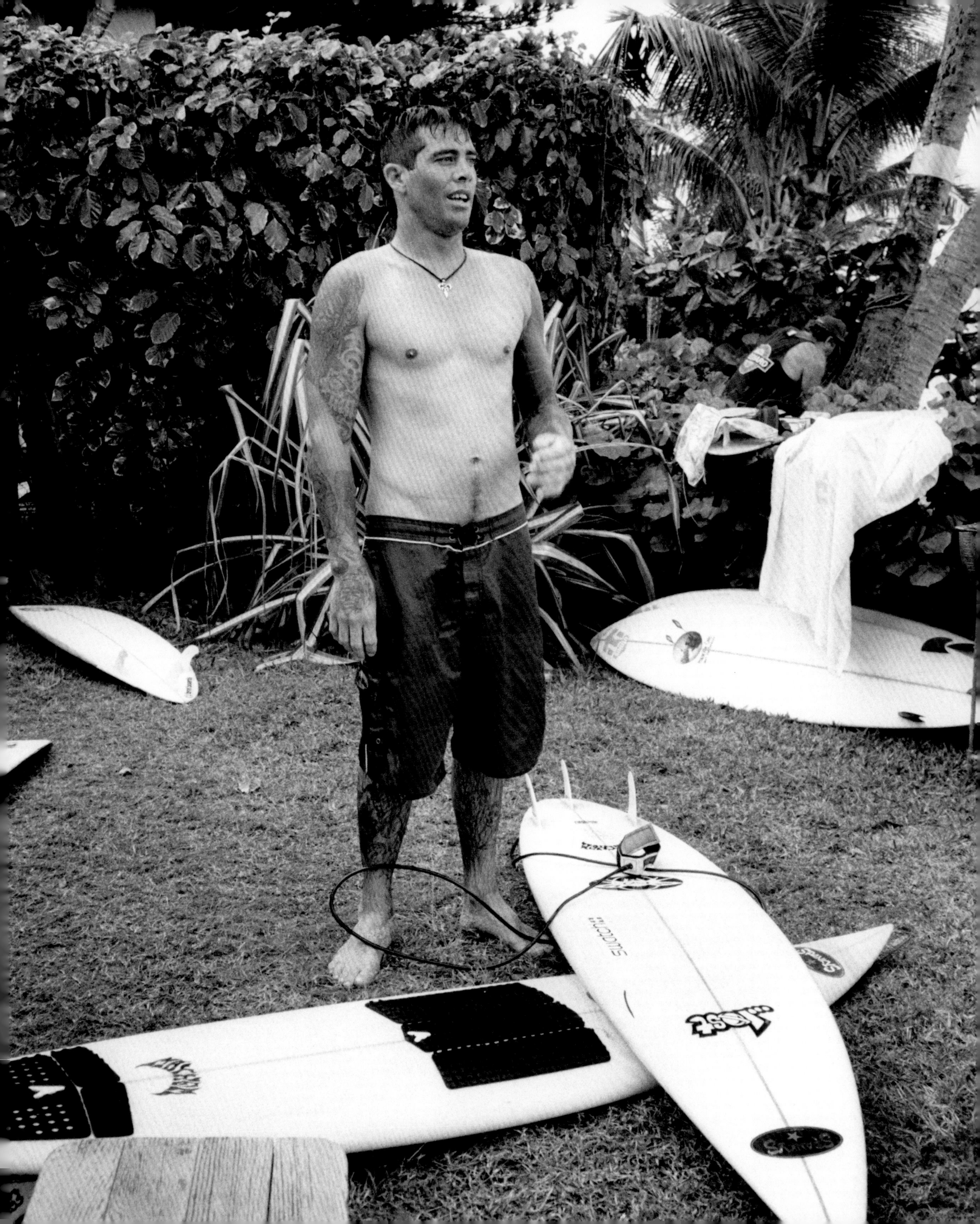

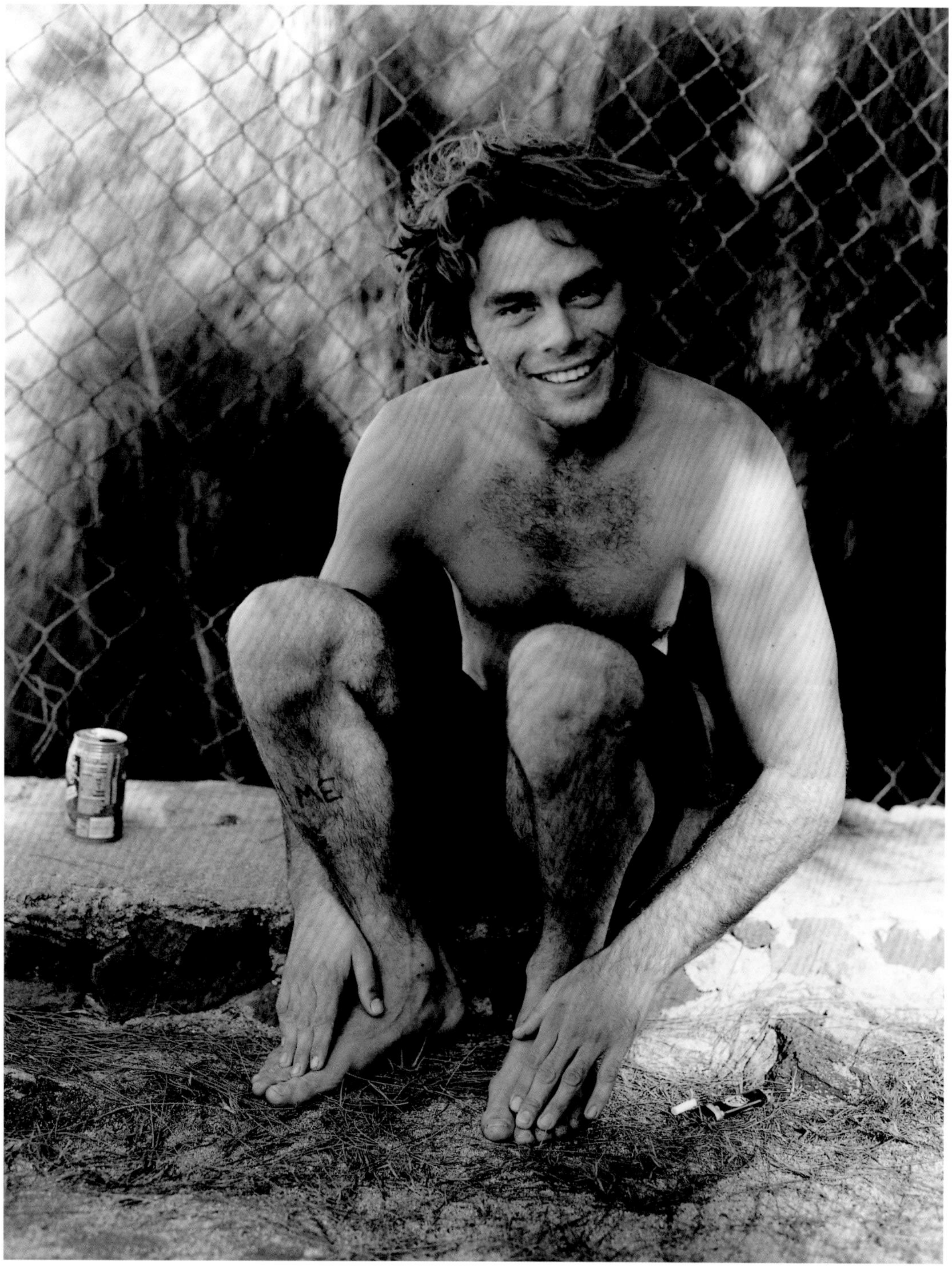

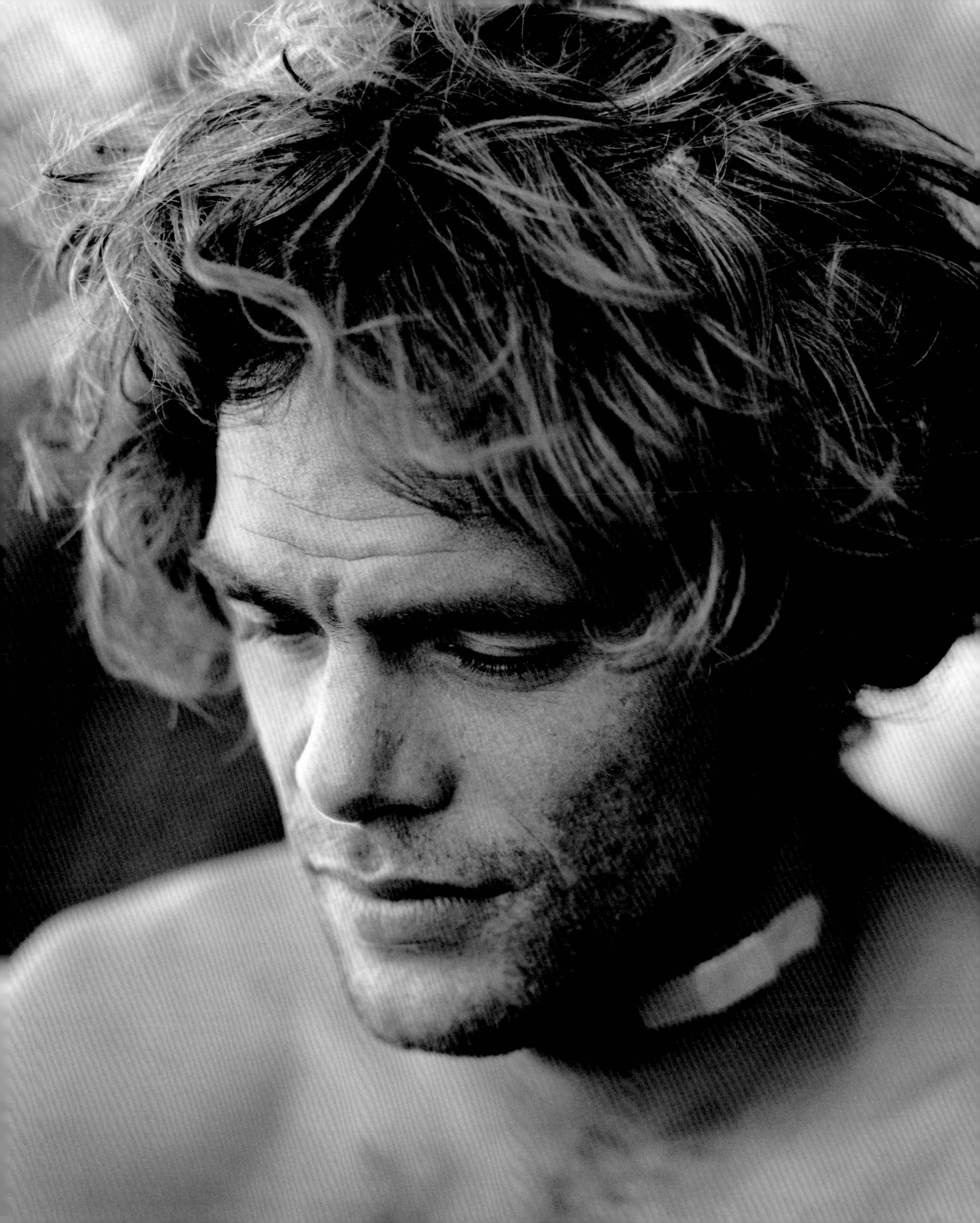

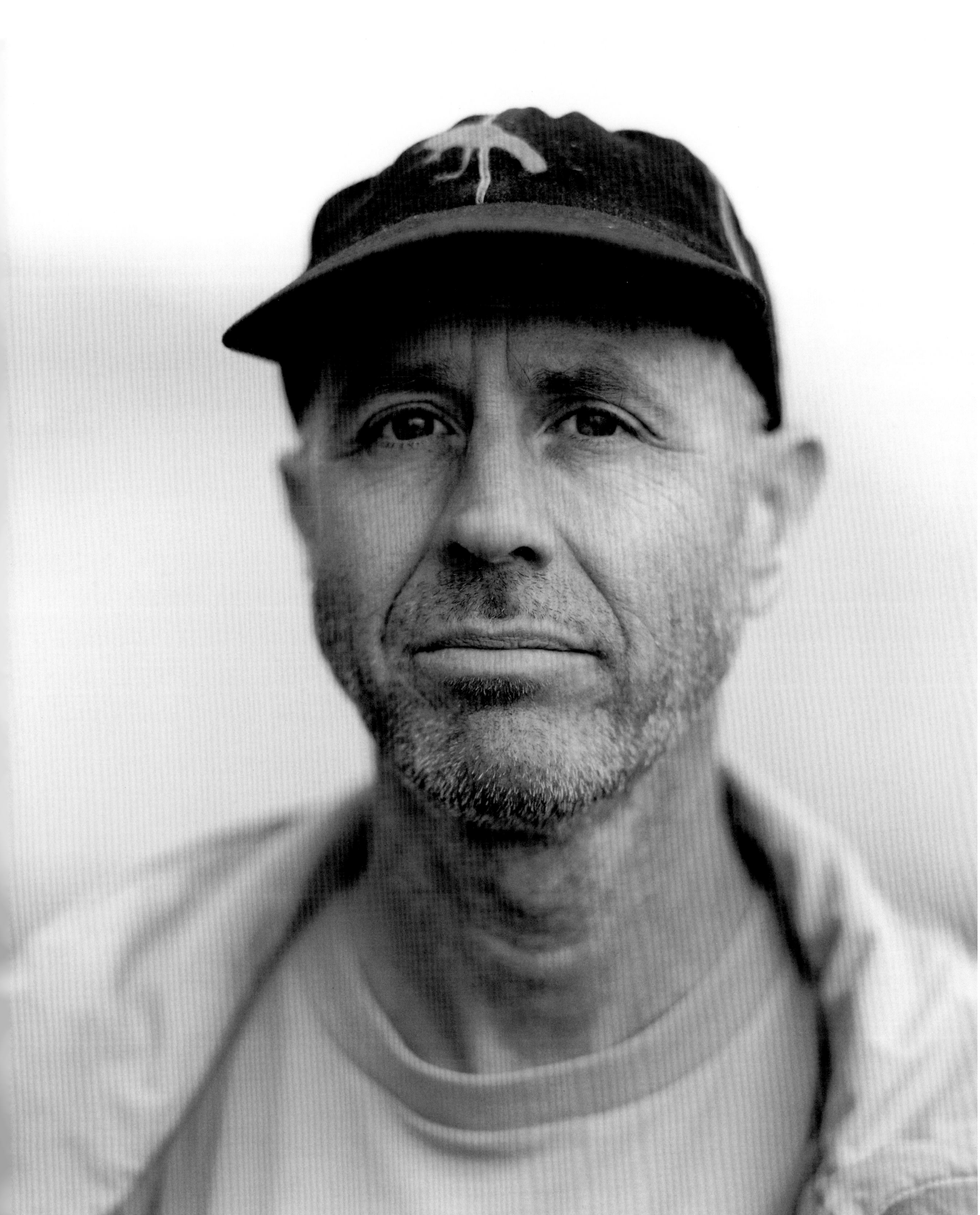

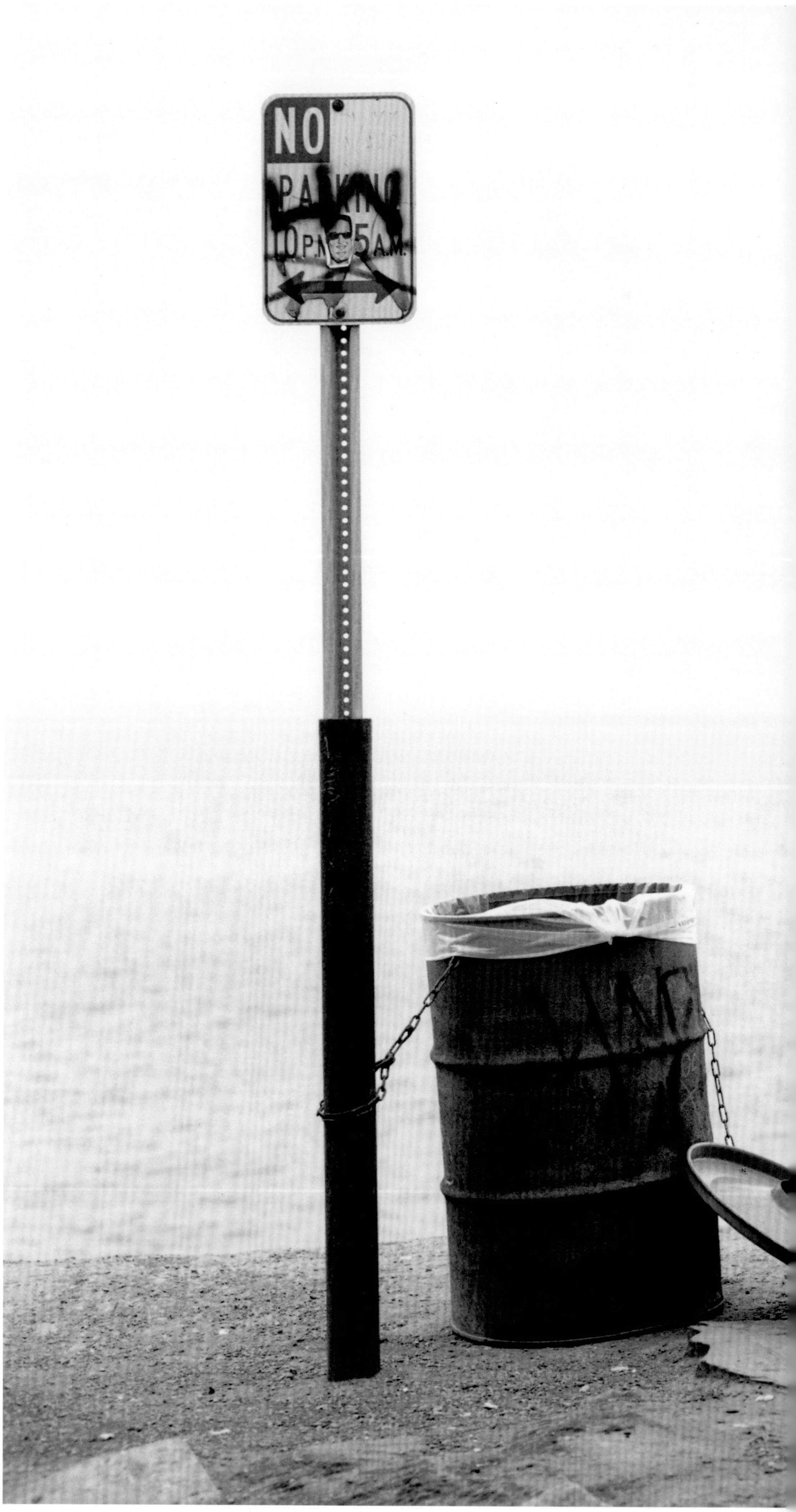
NO

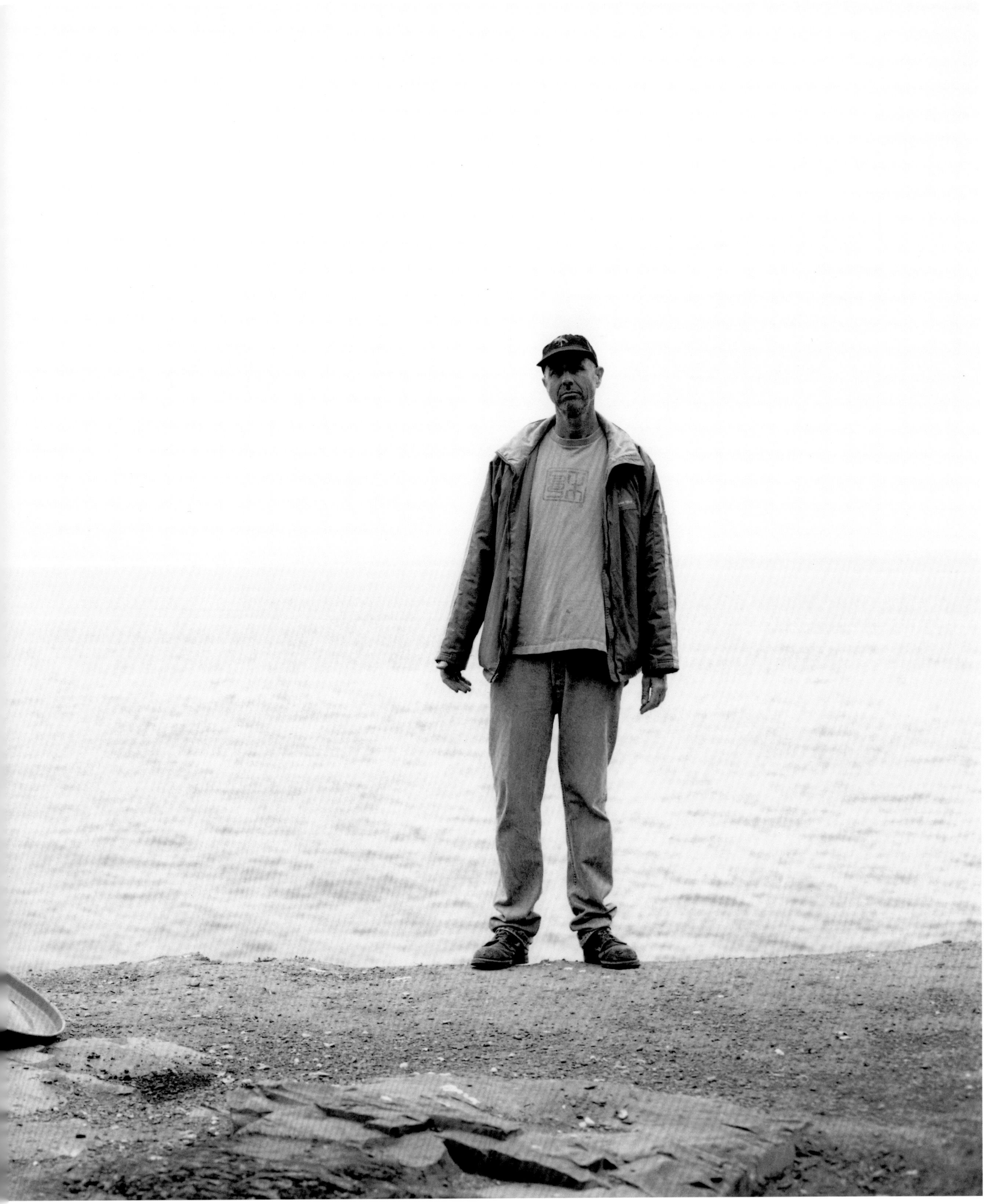

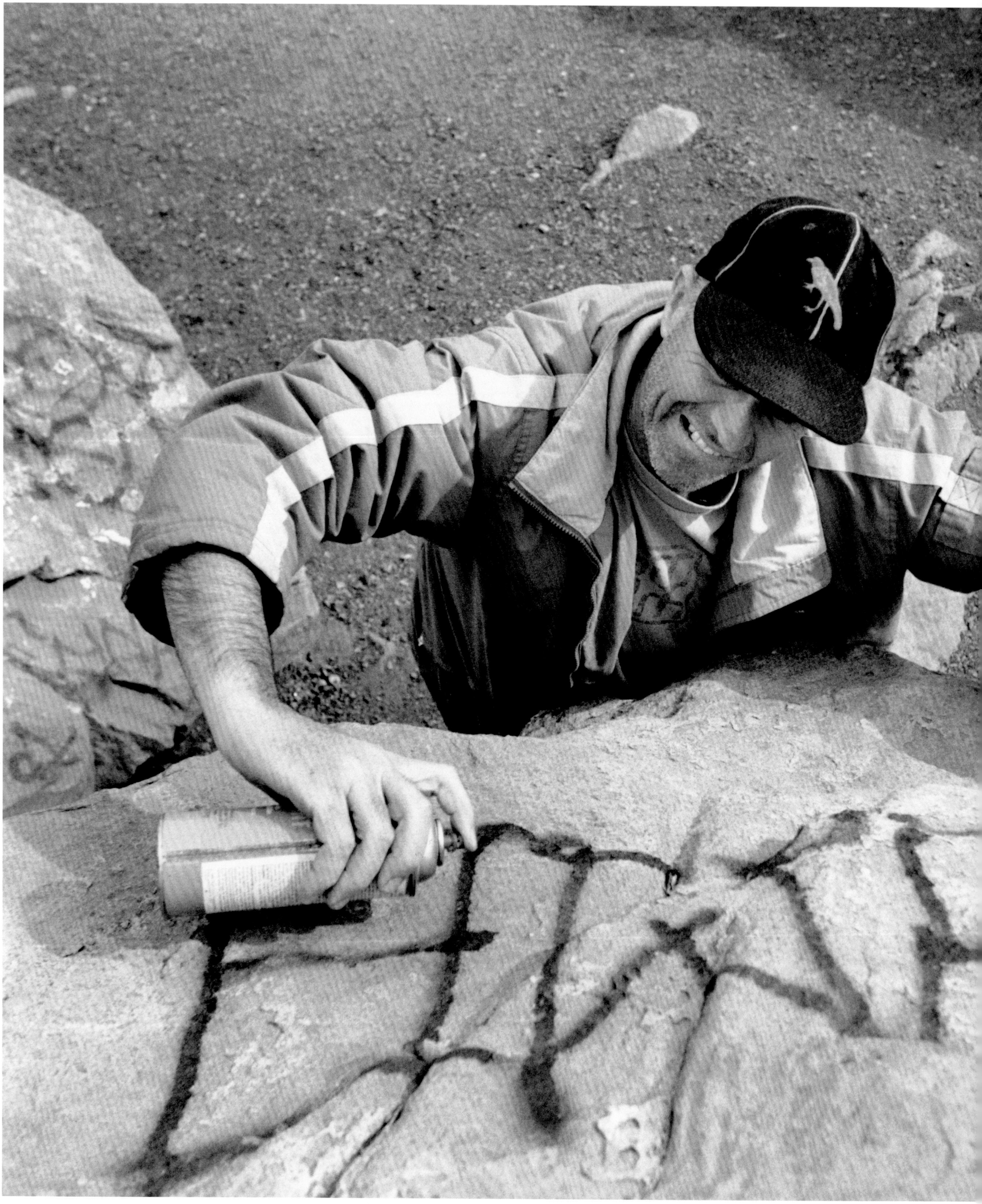

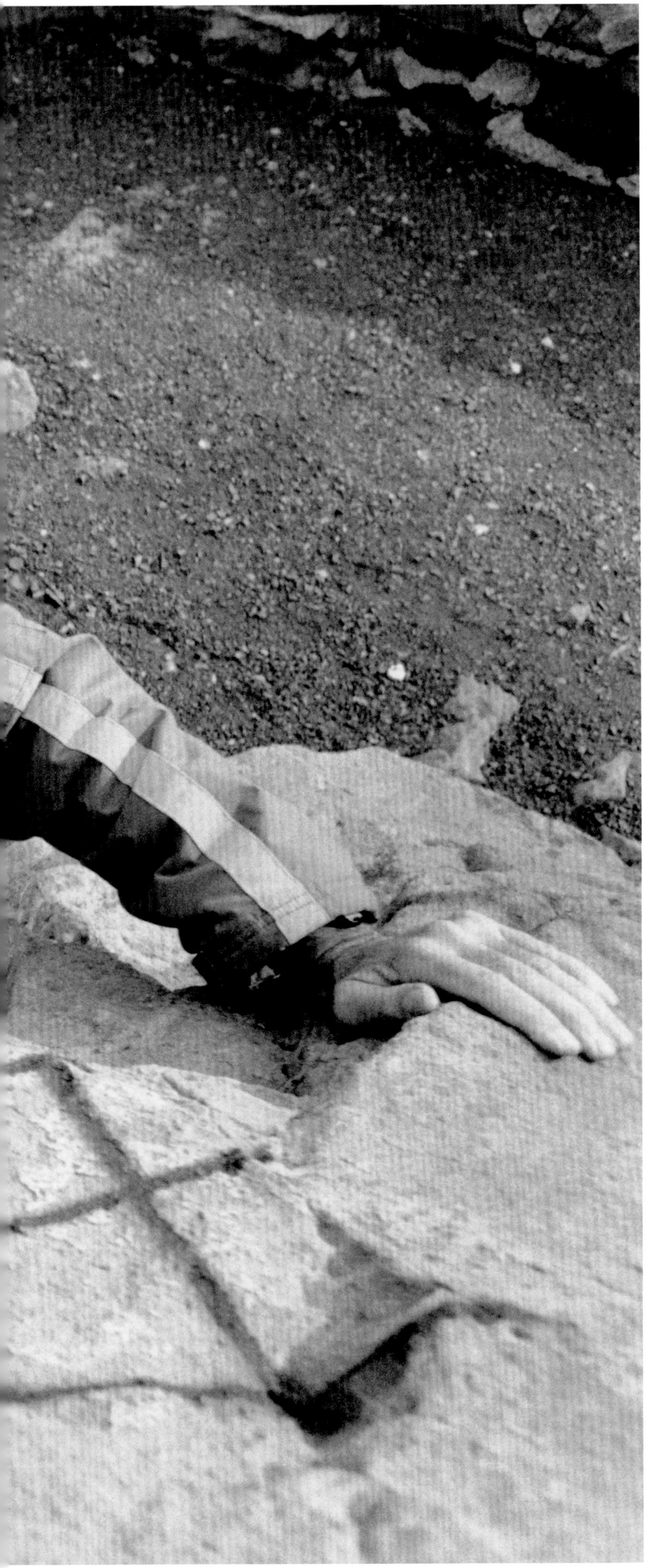

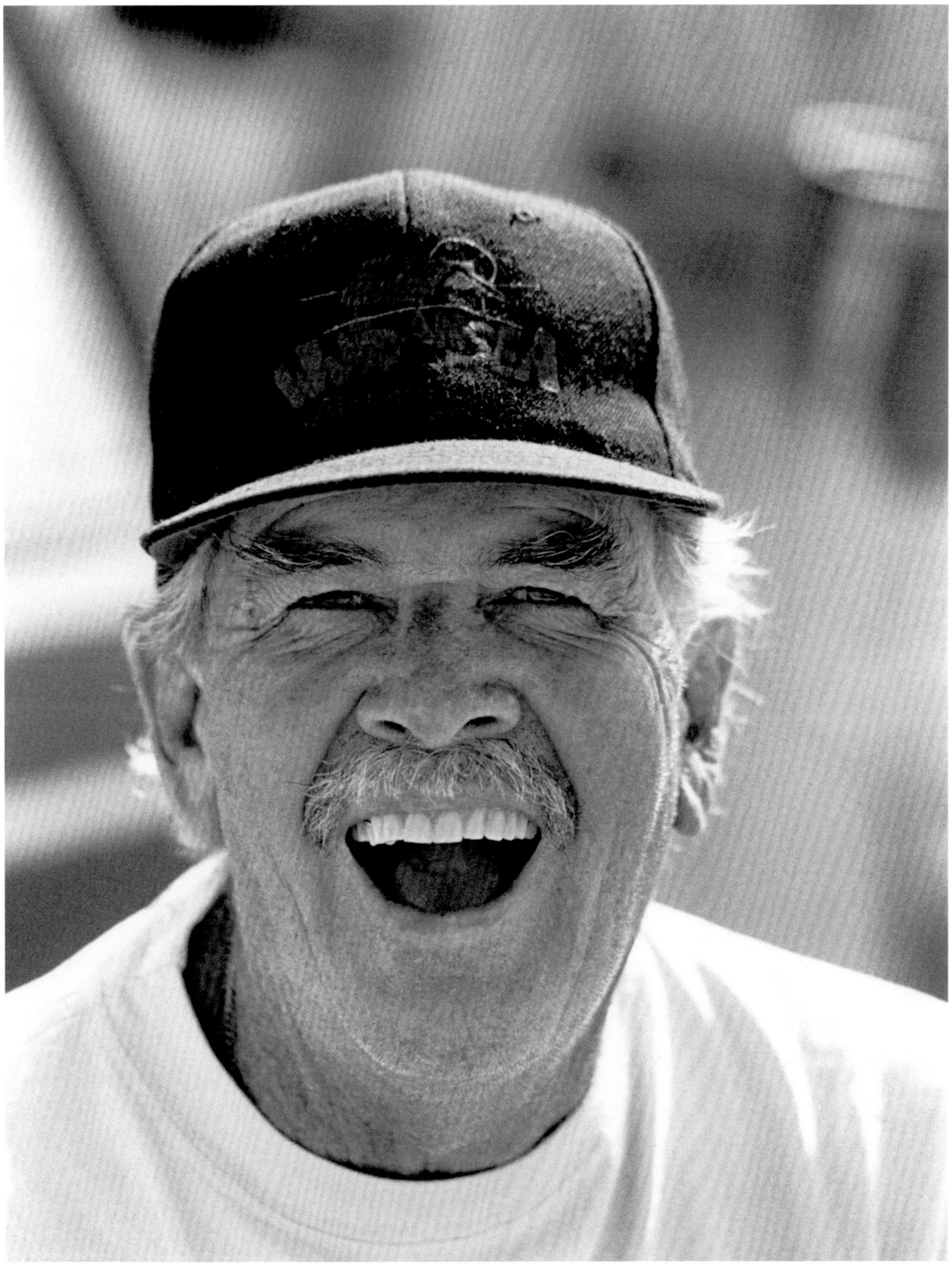

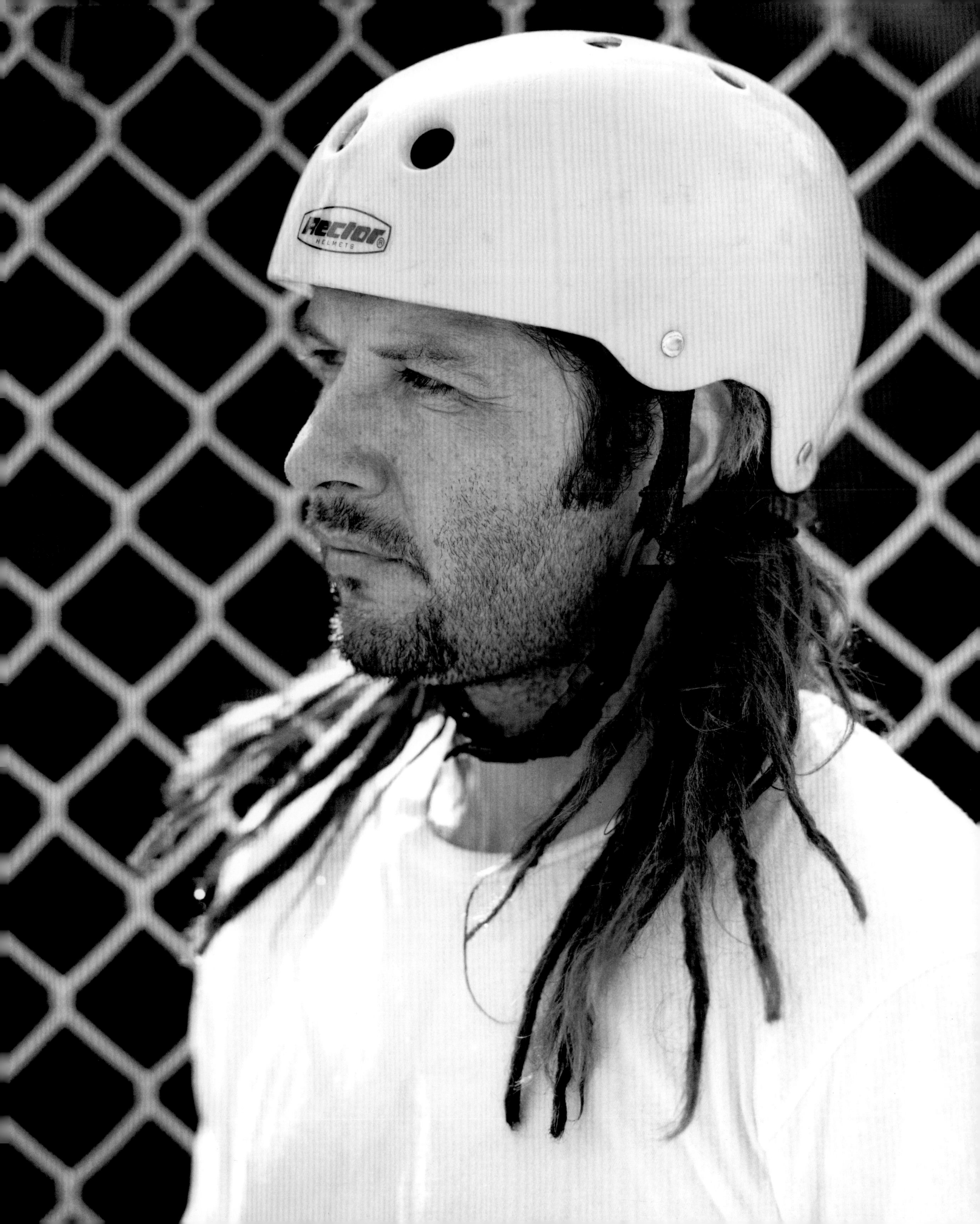

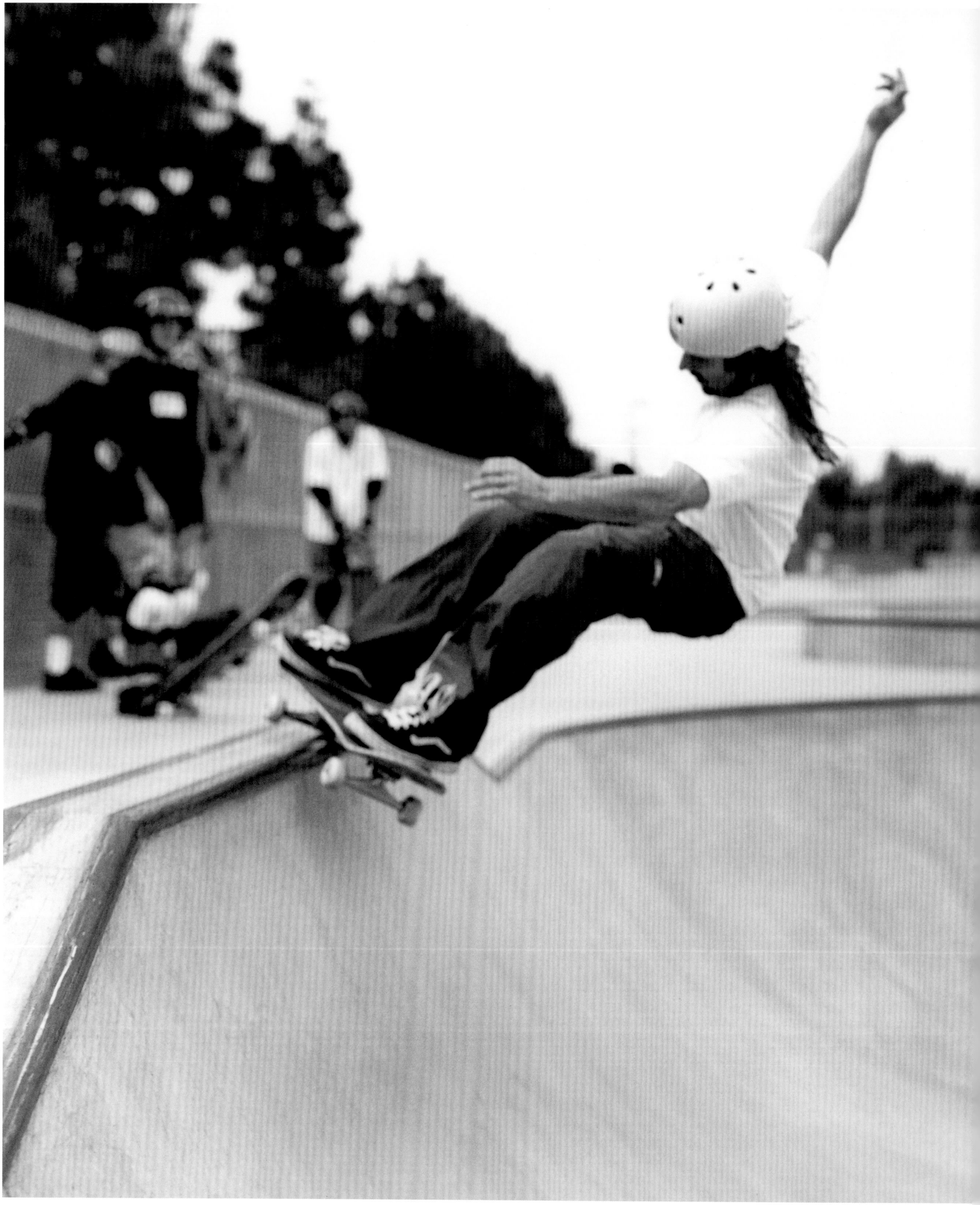

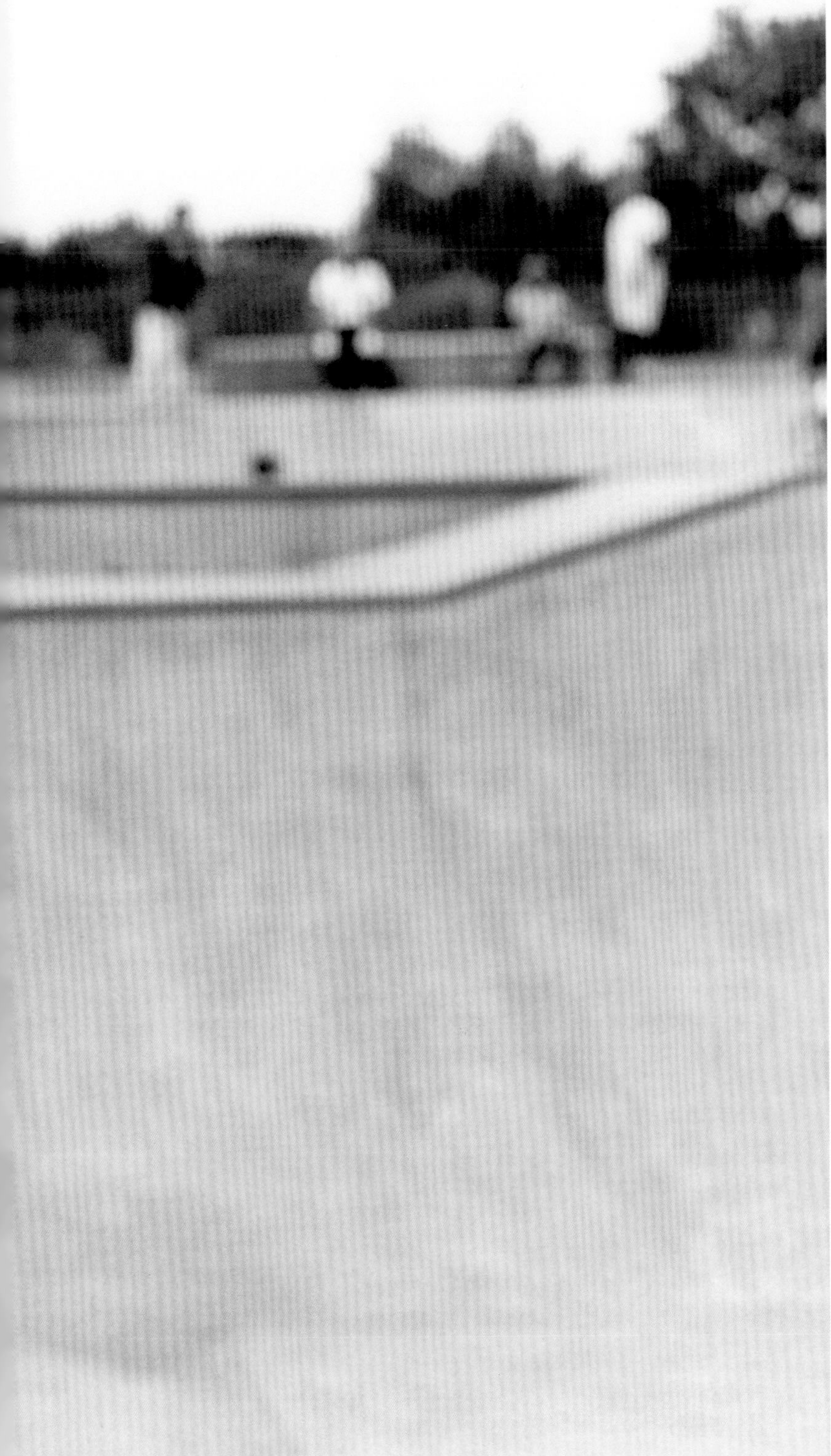

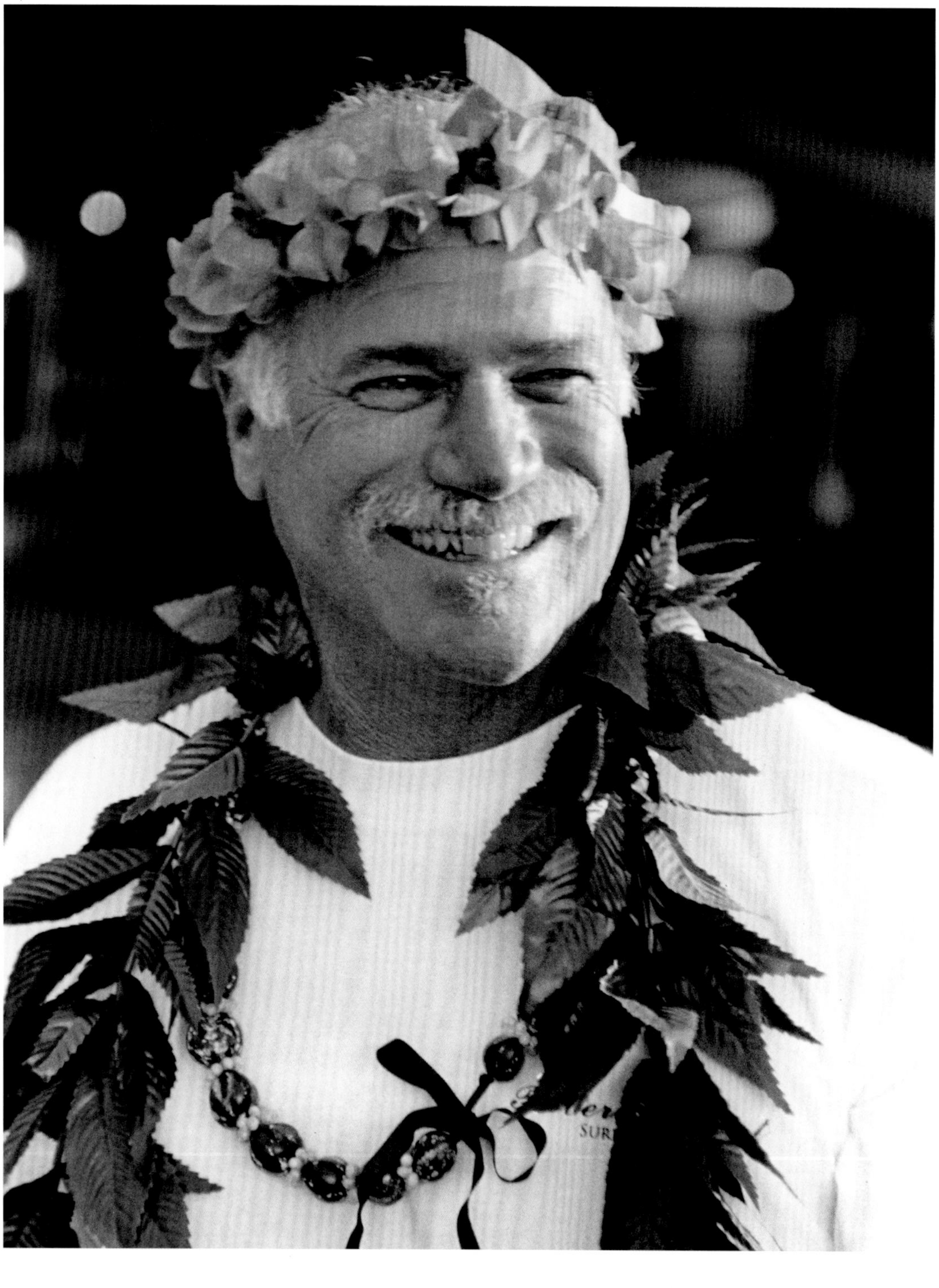

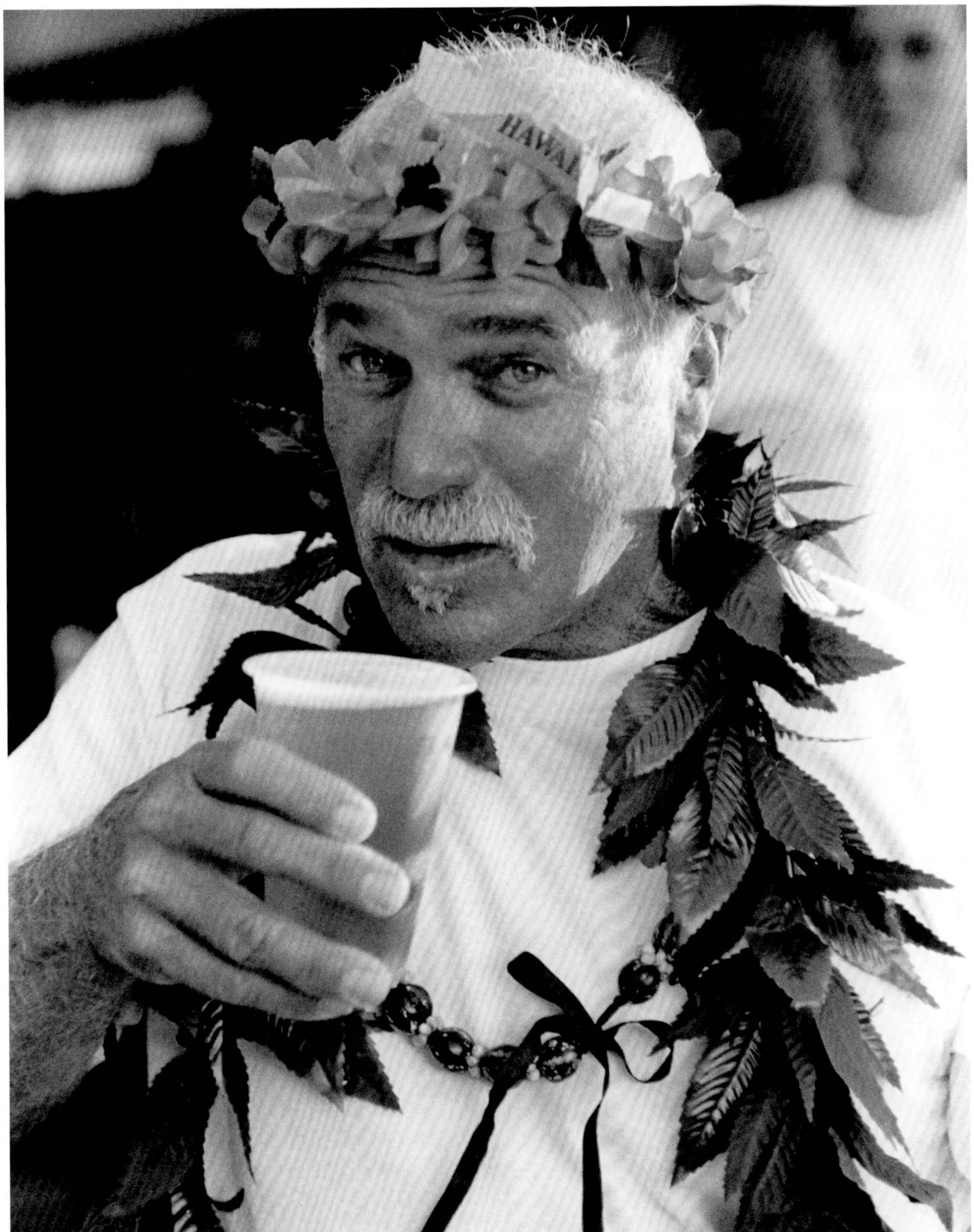
HAWAI

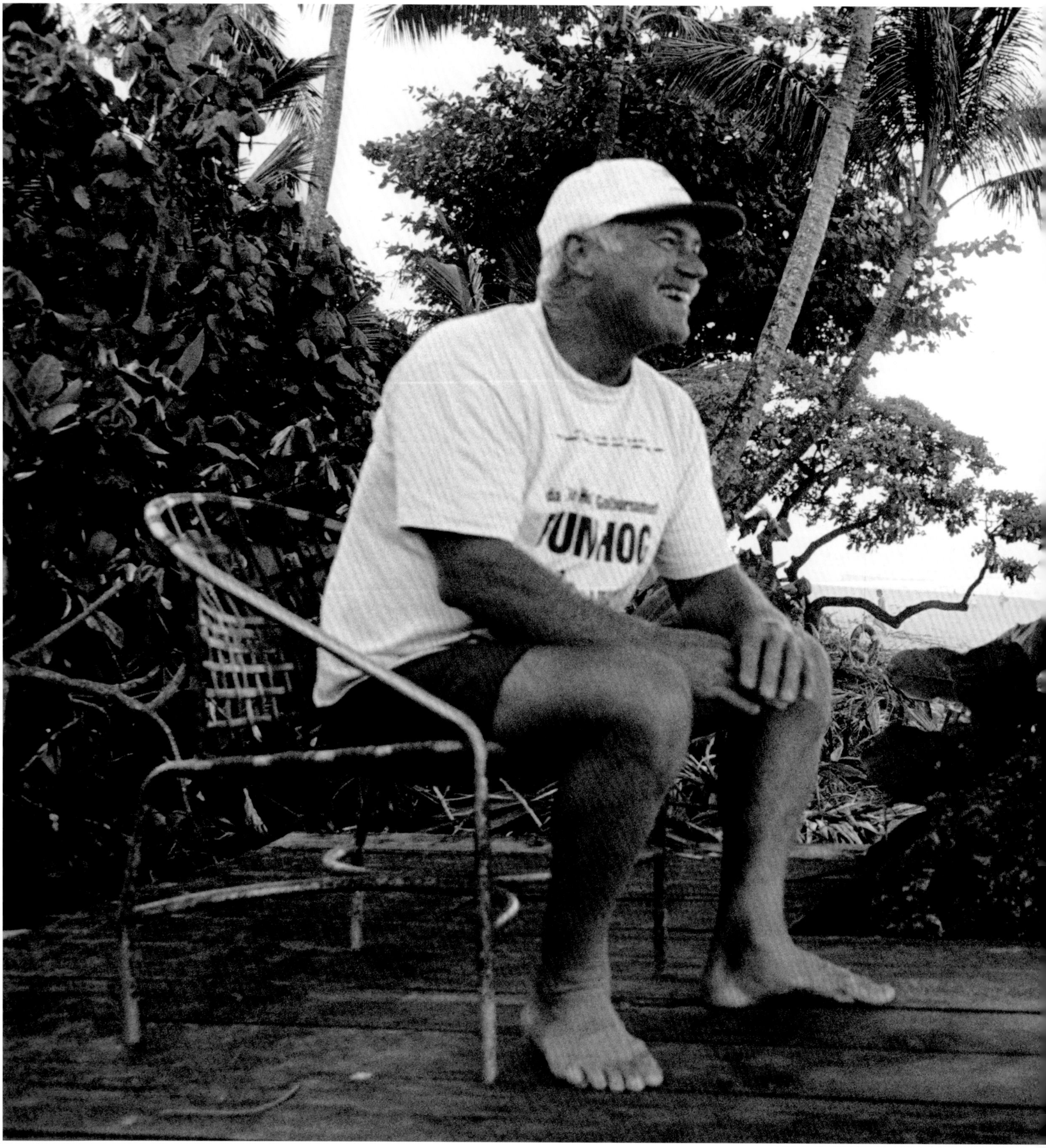

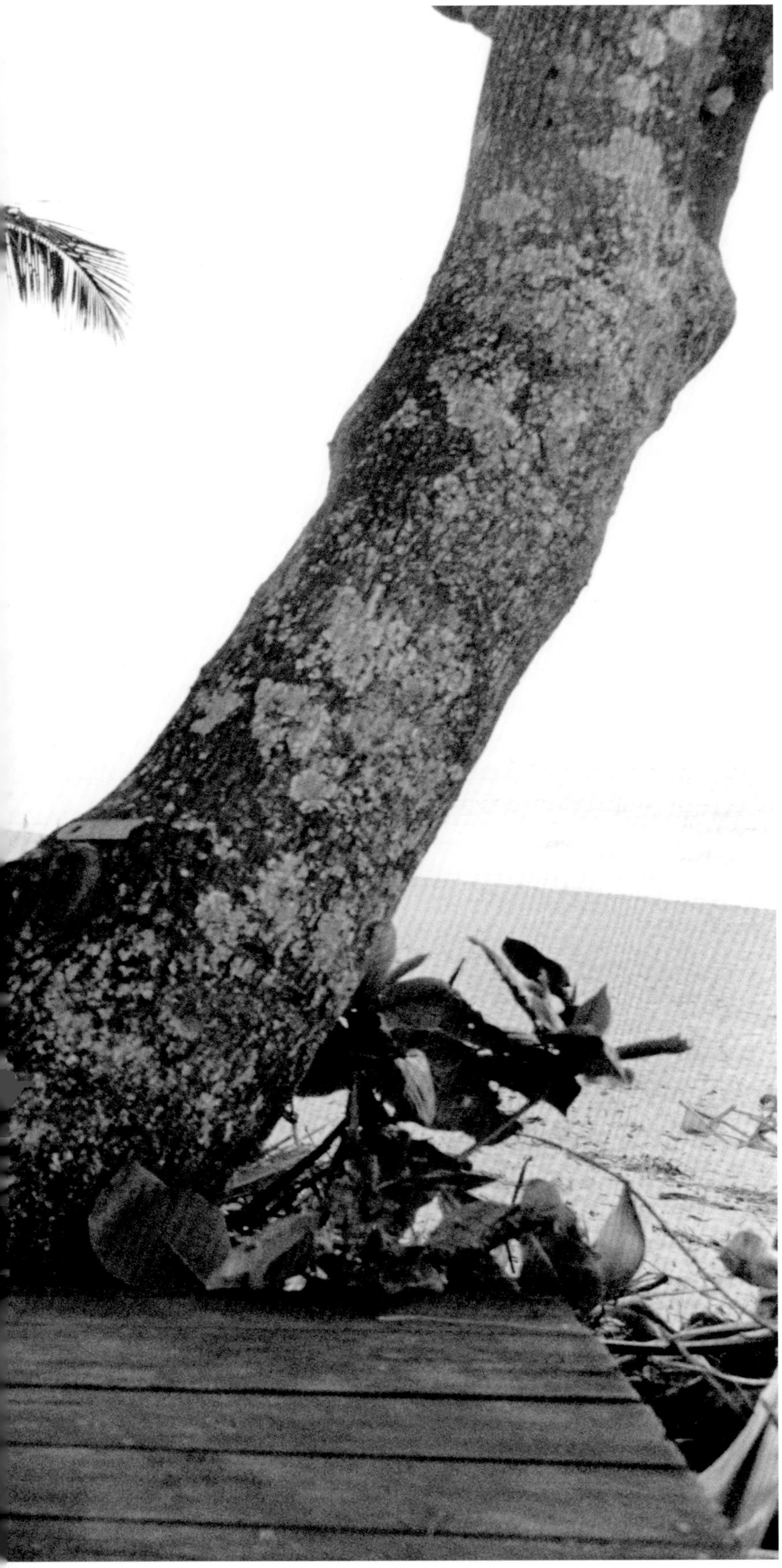

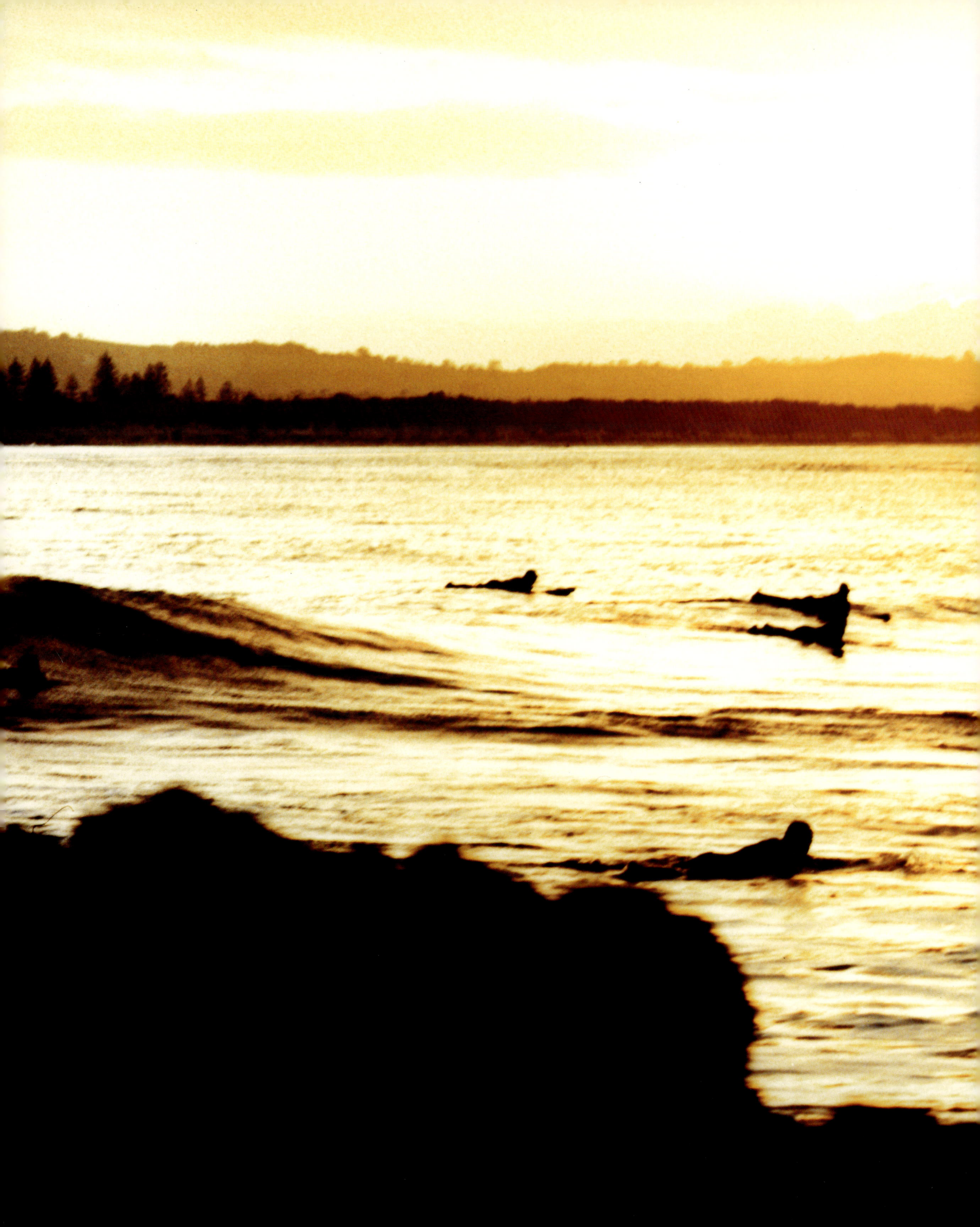

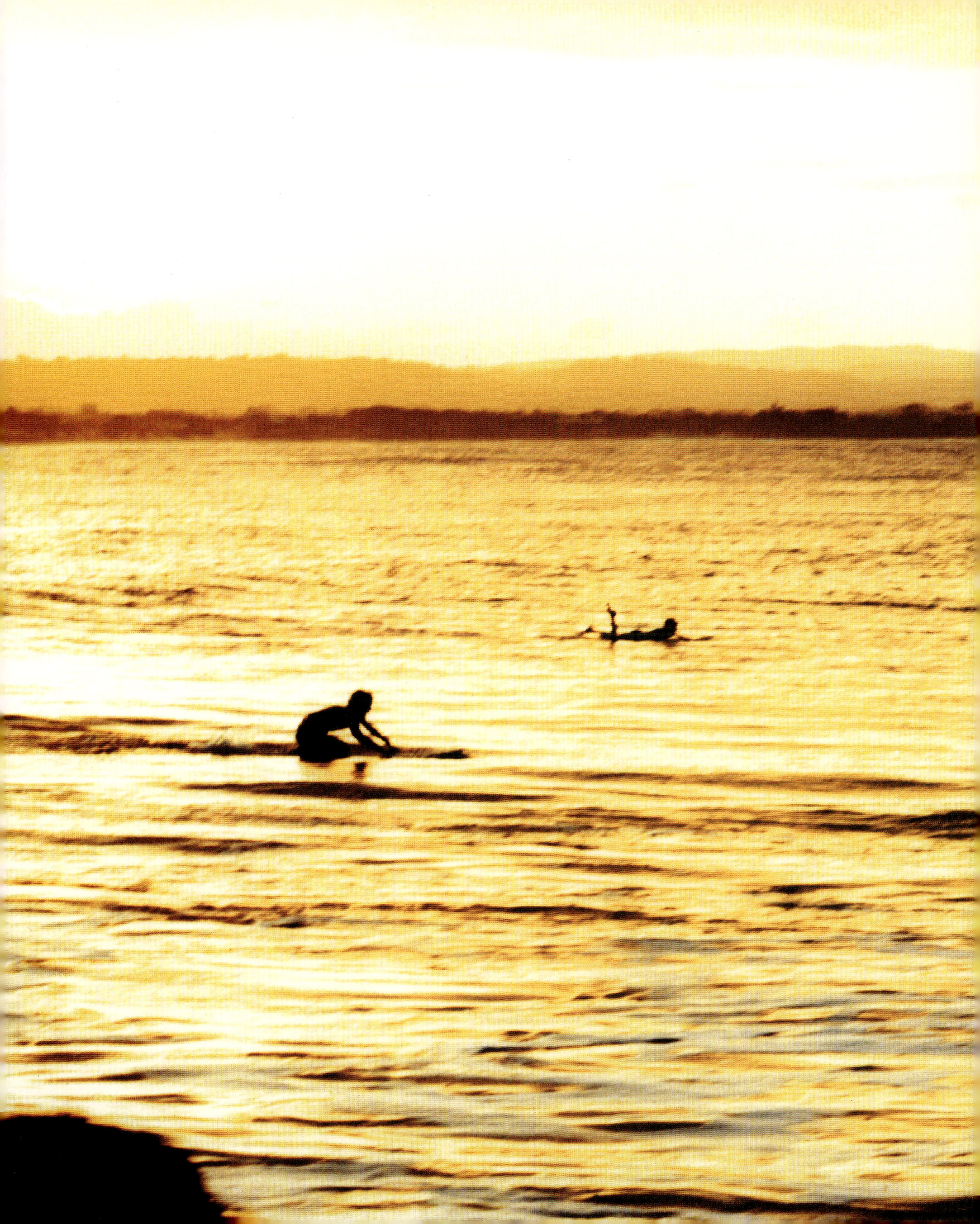

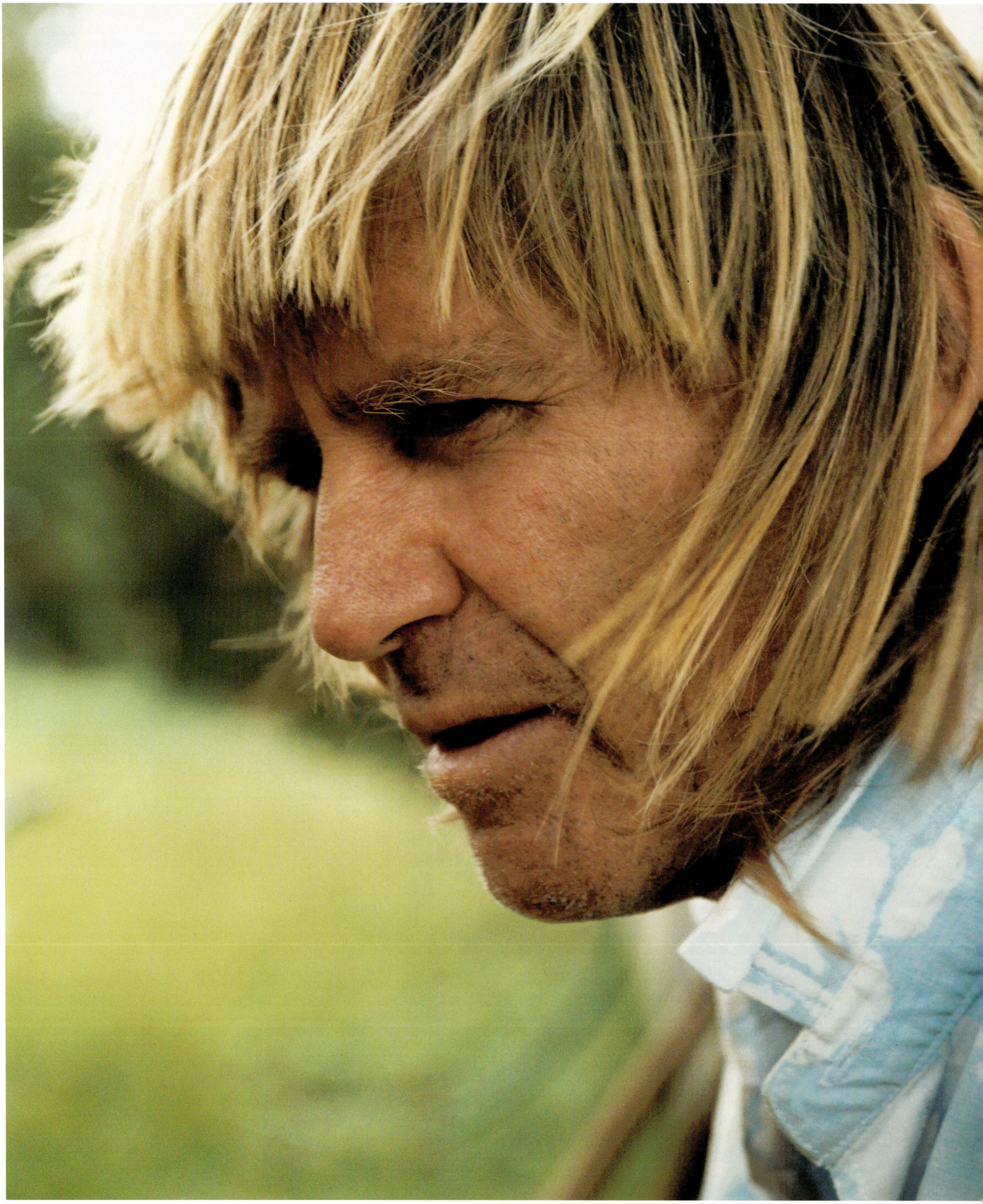

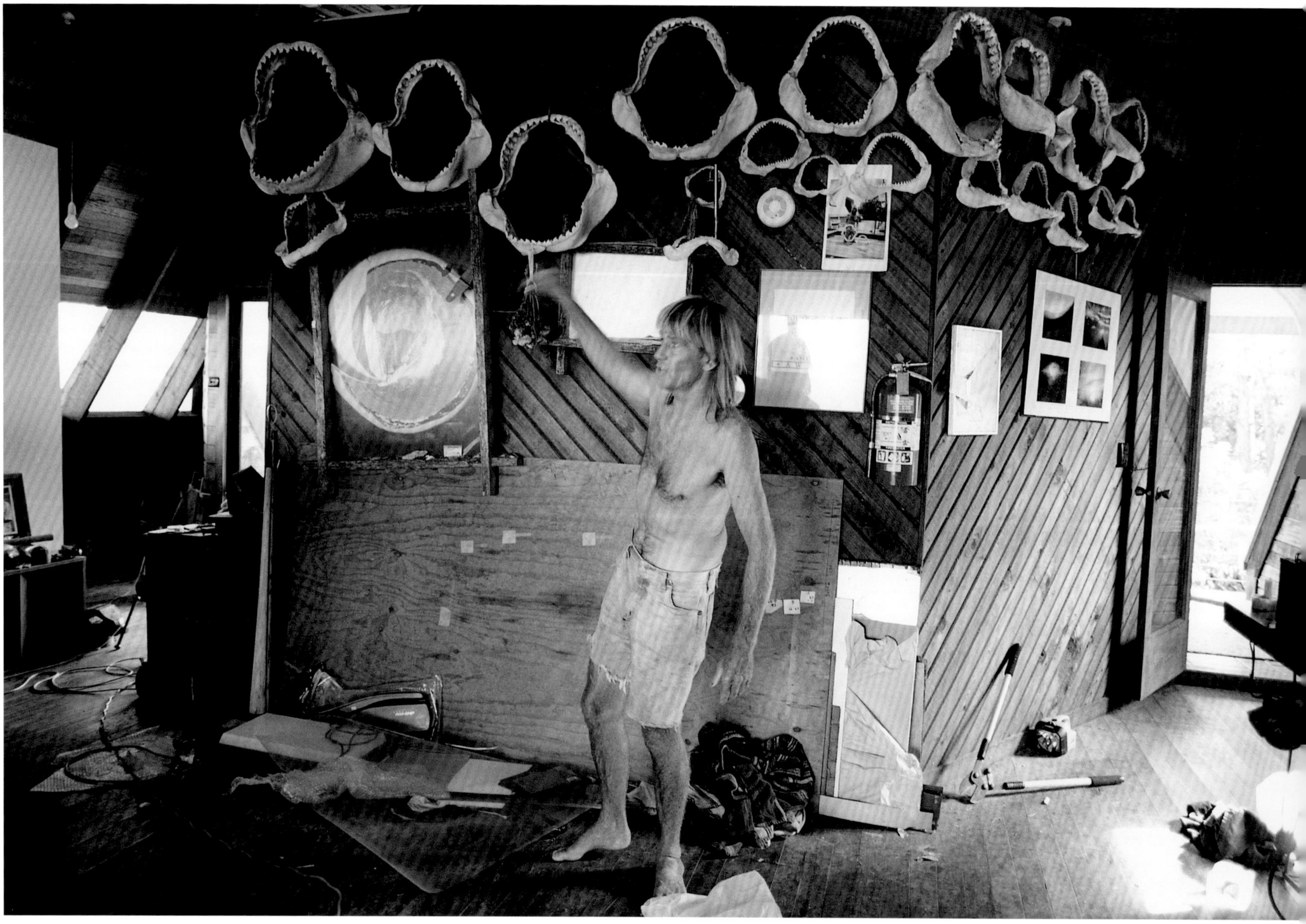

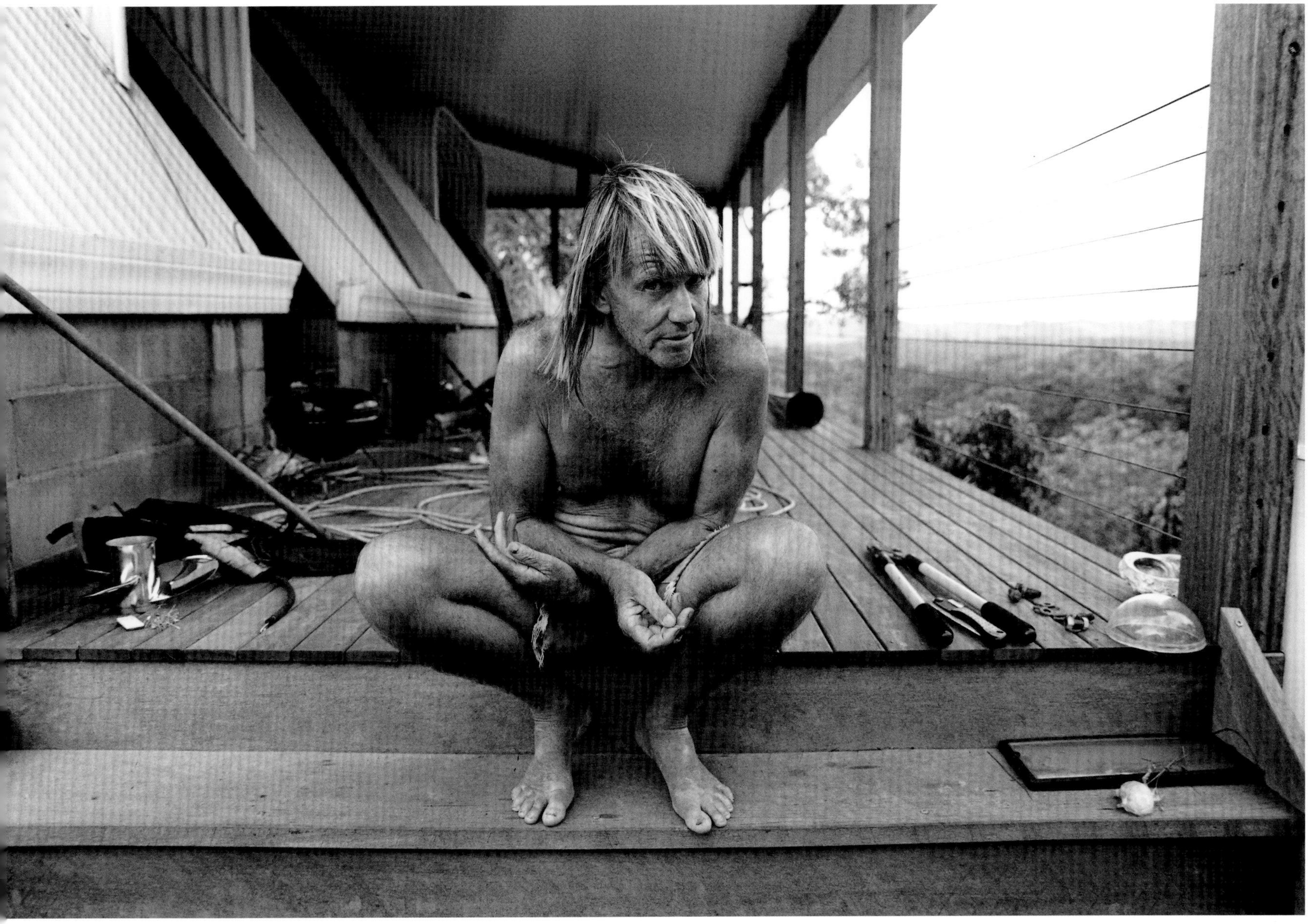

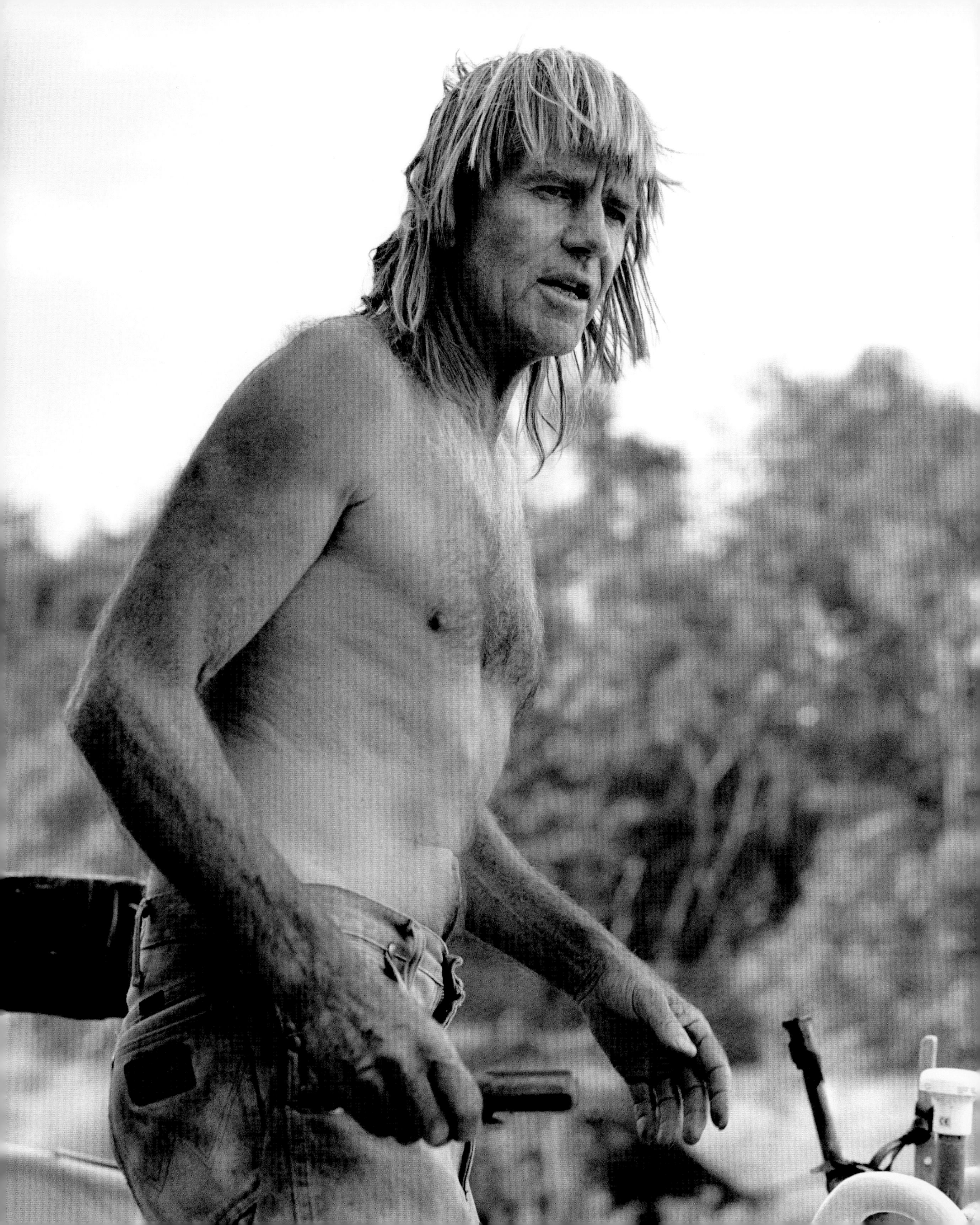

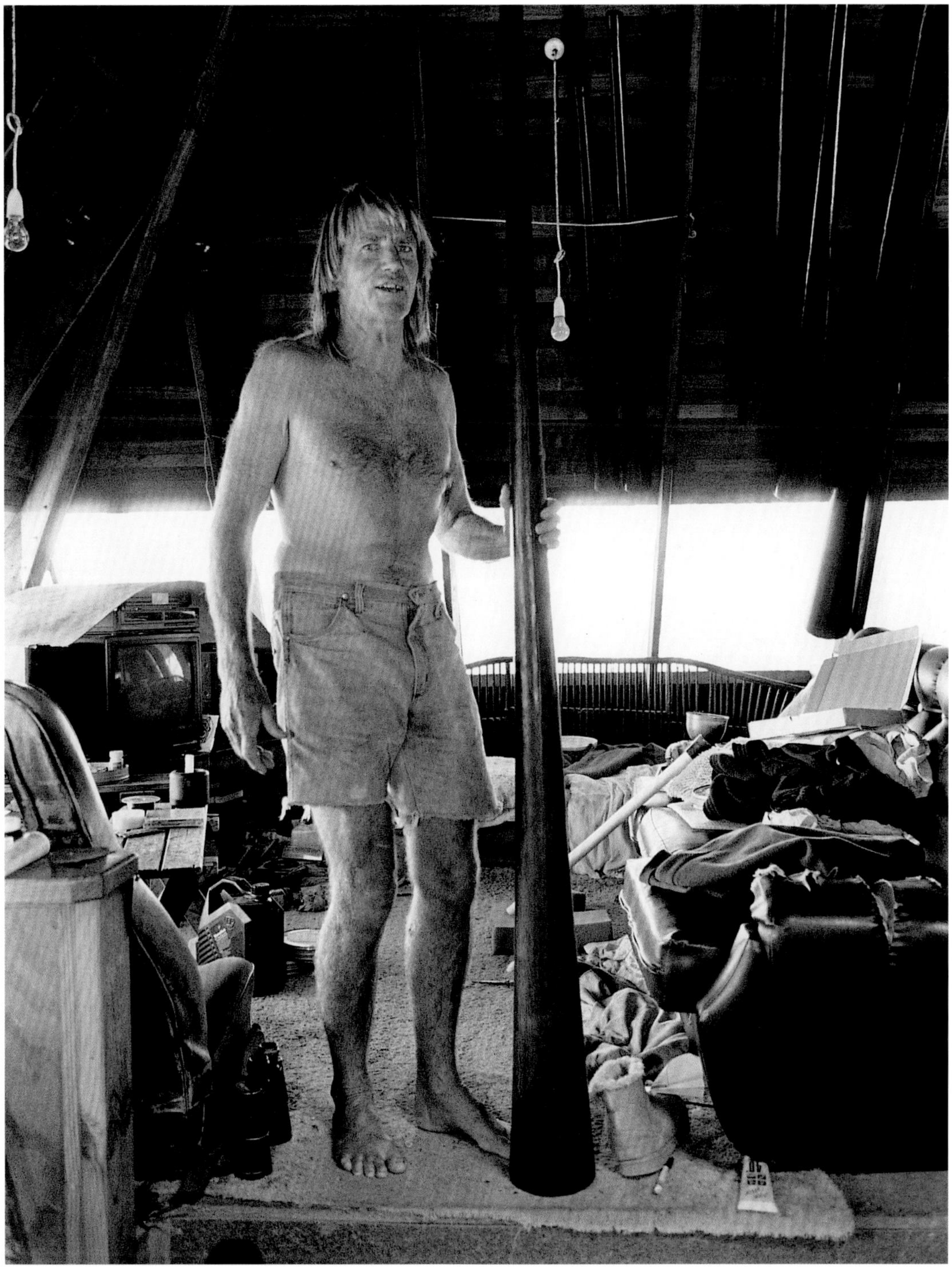

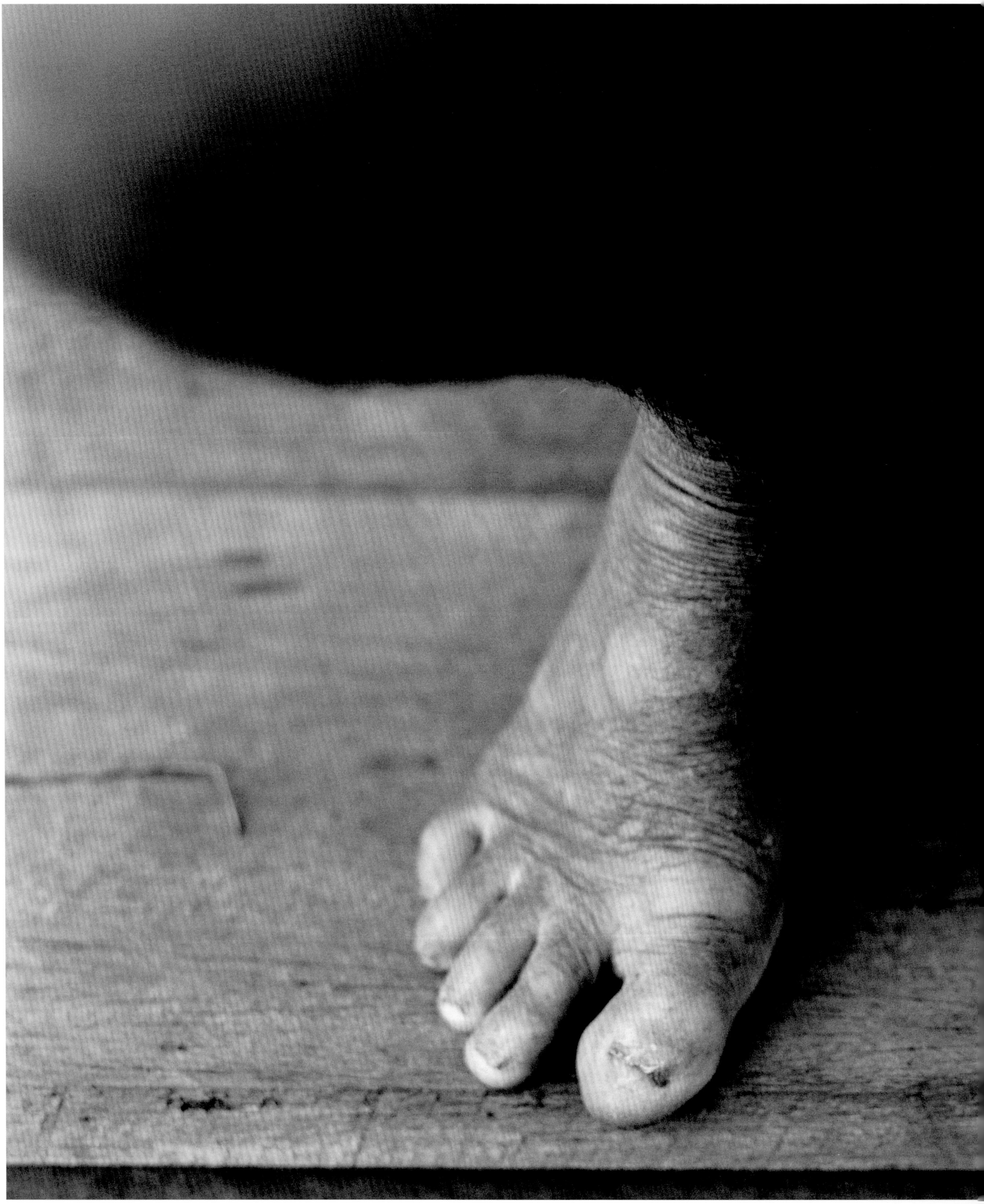

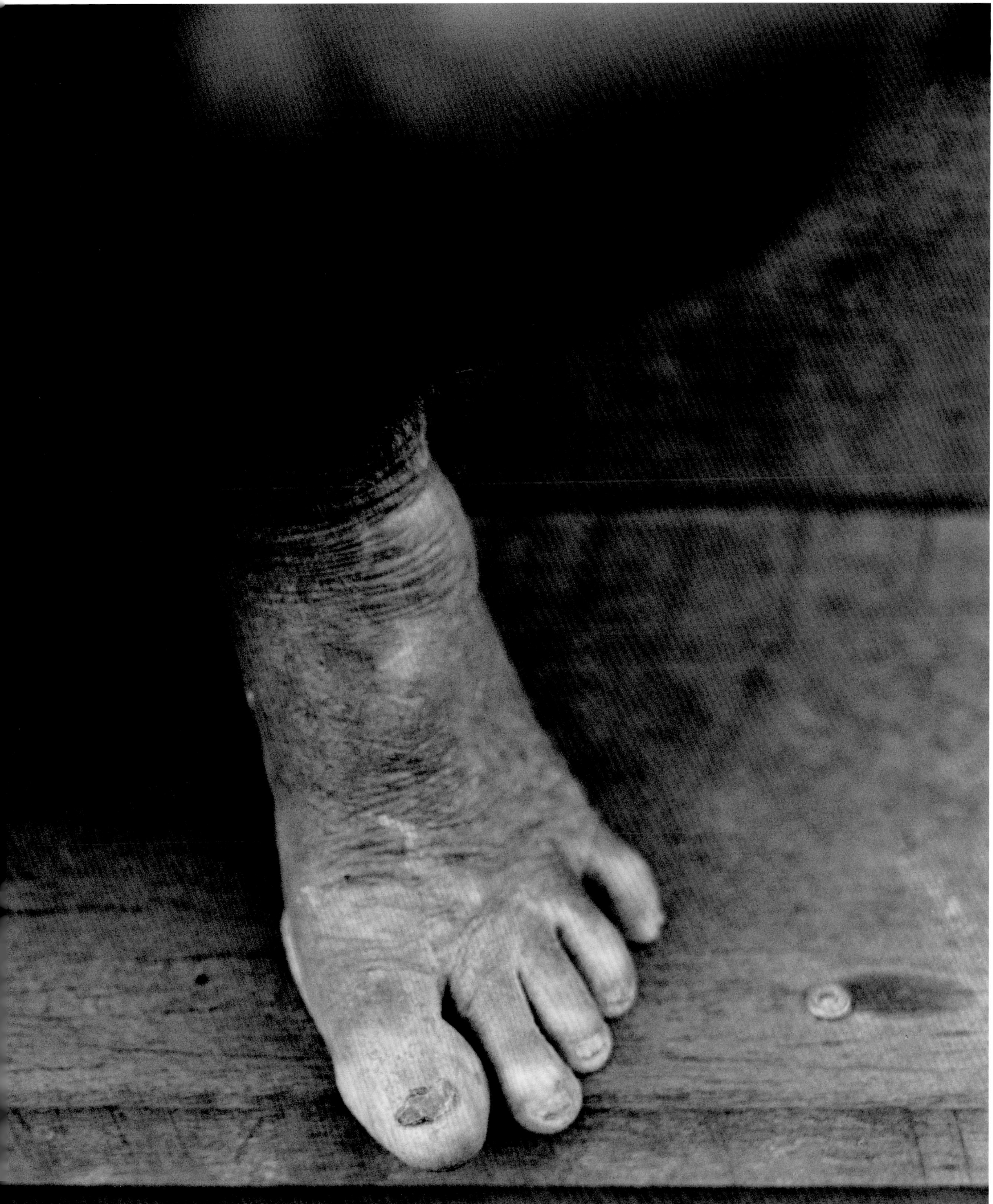

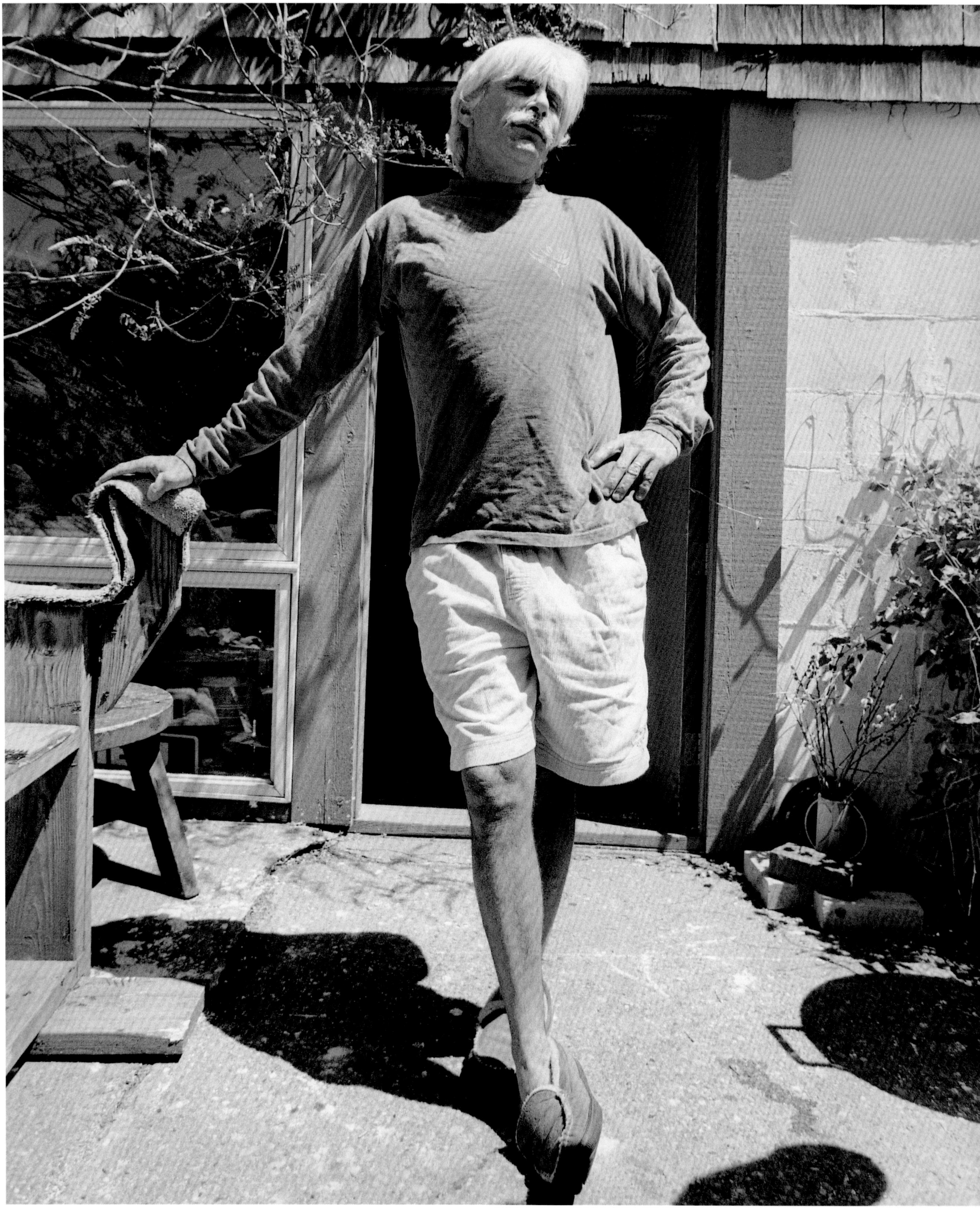

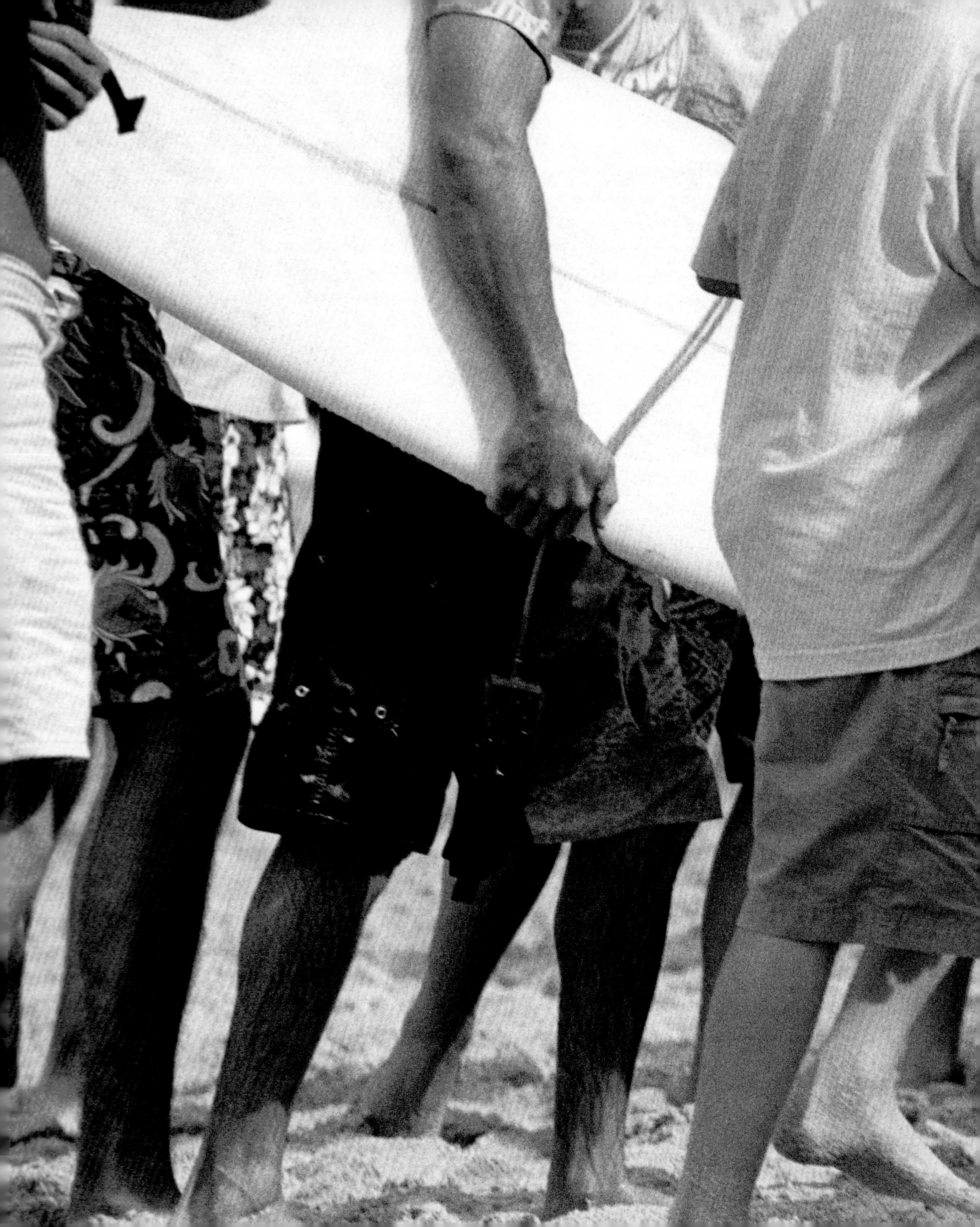

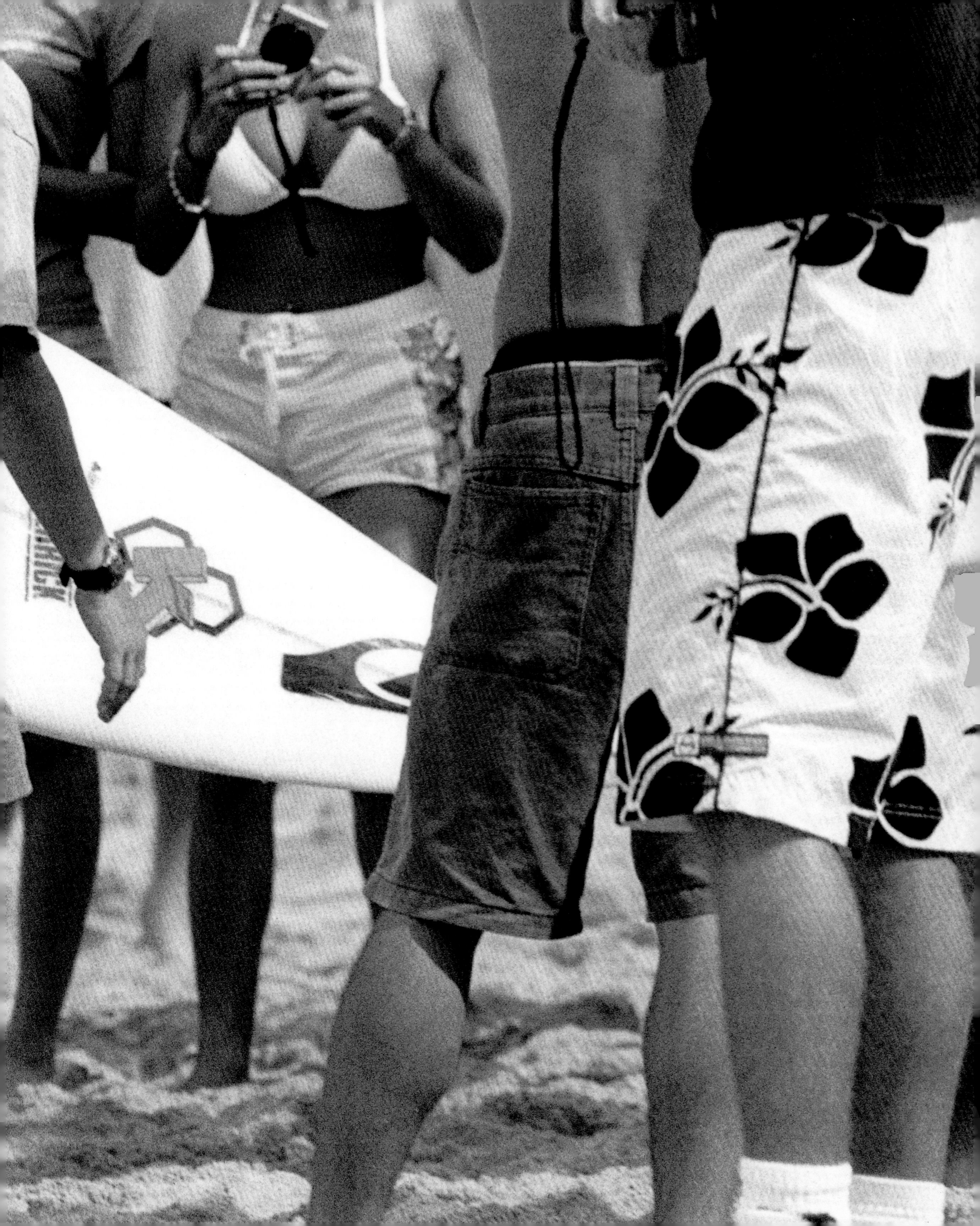

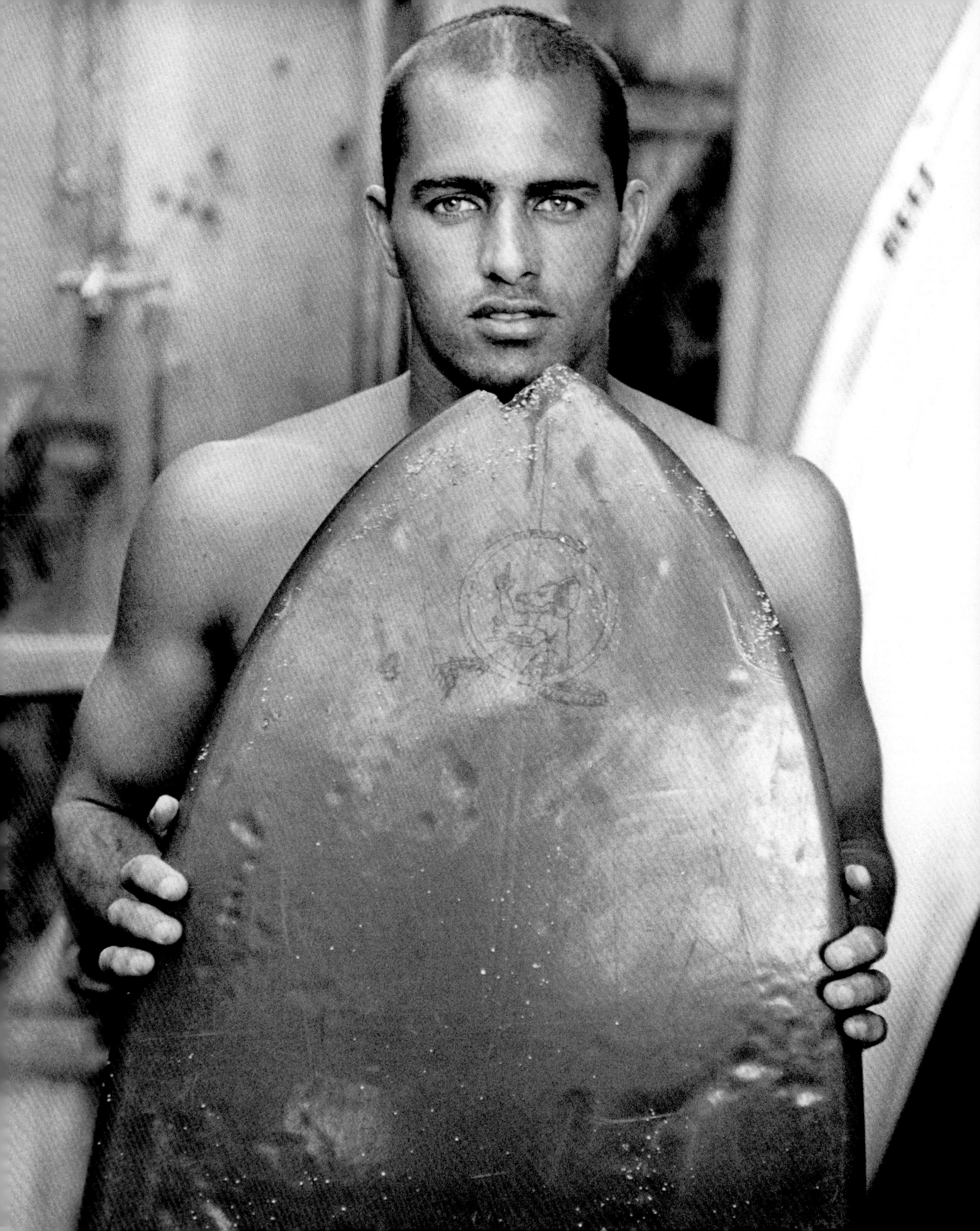

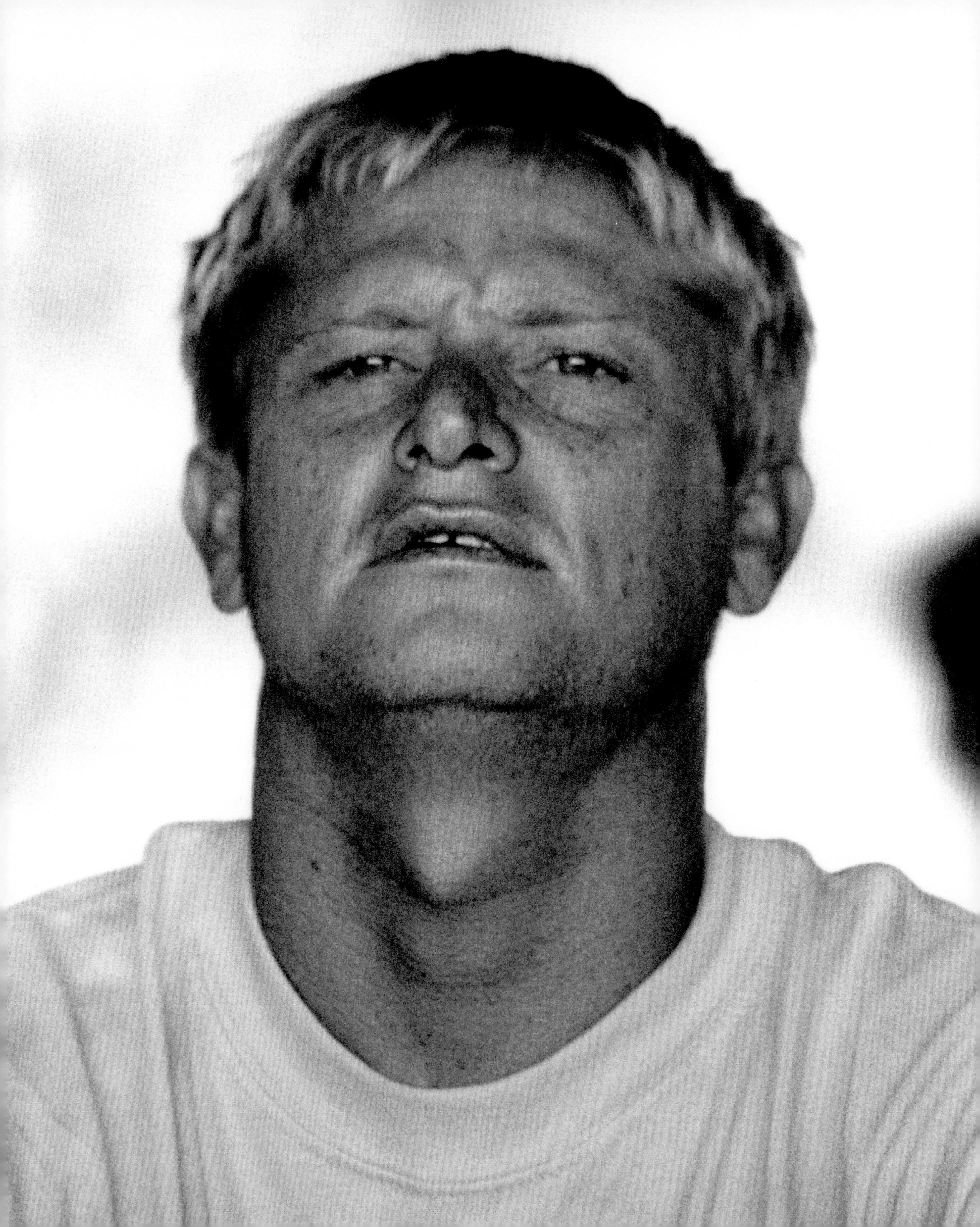

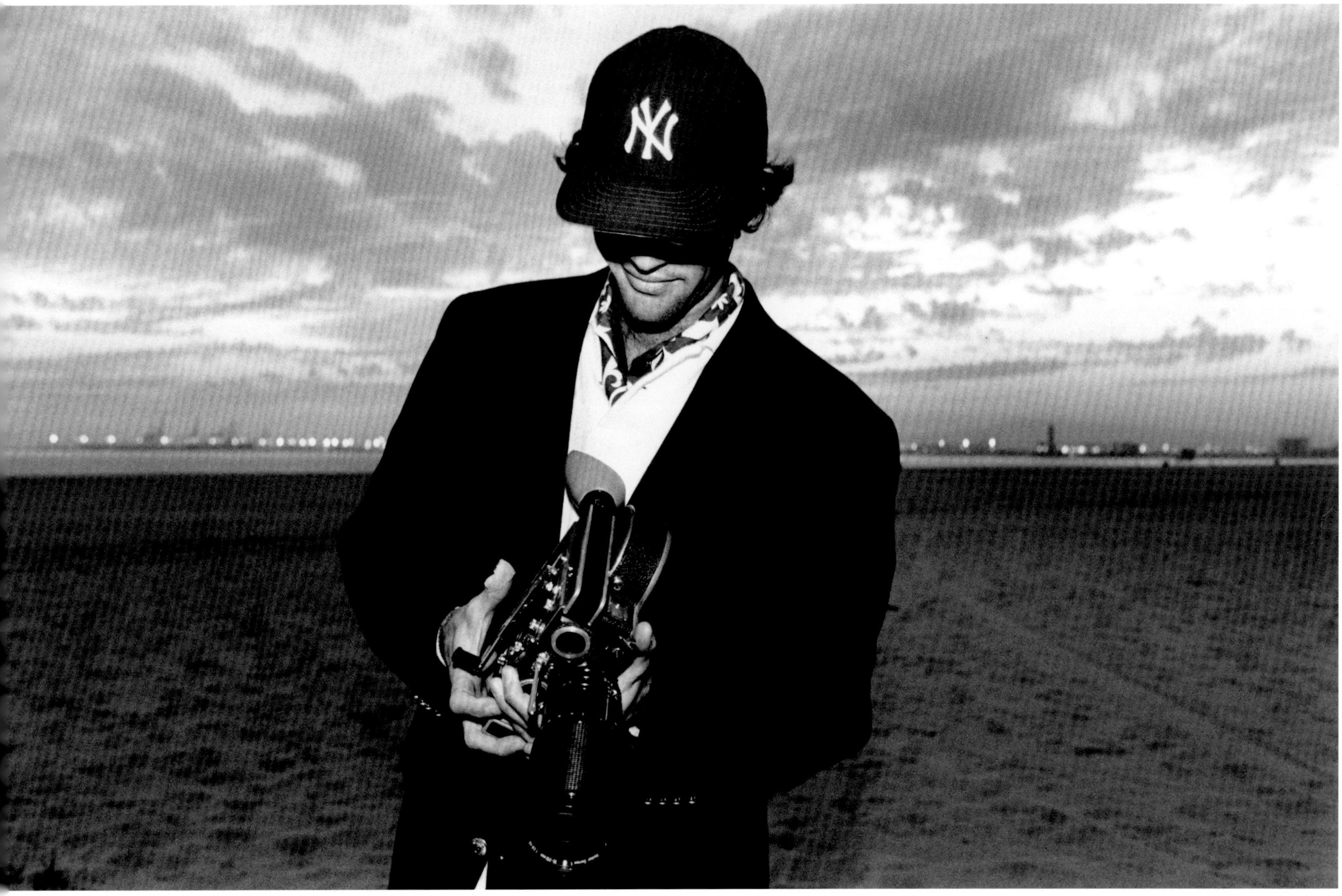
NY

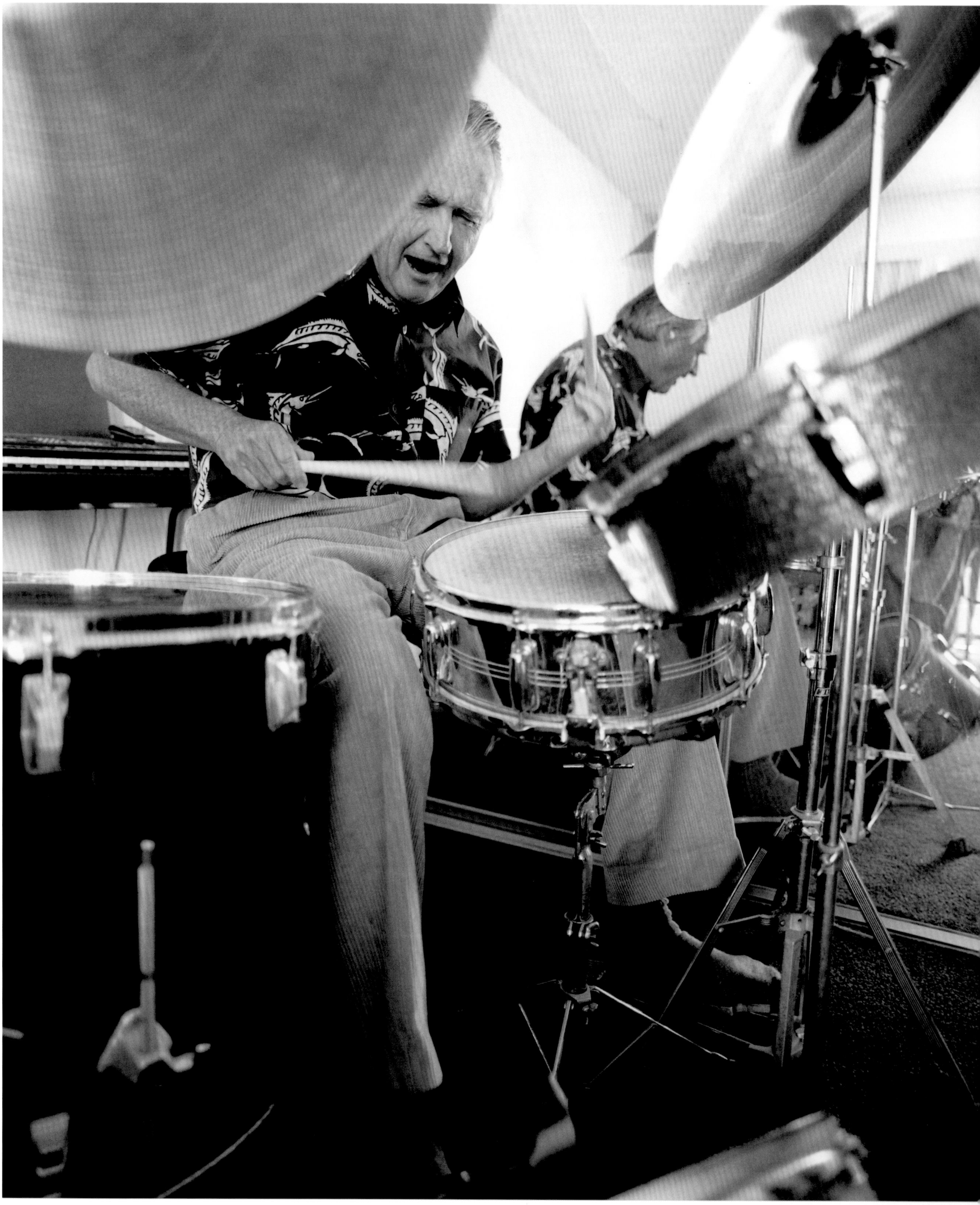

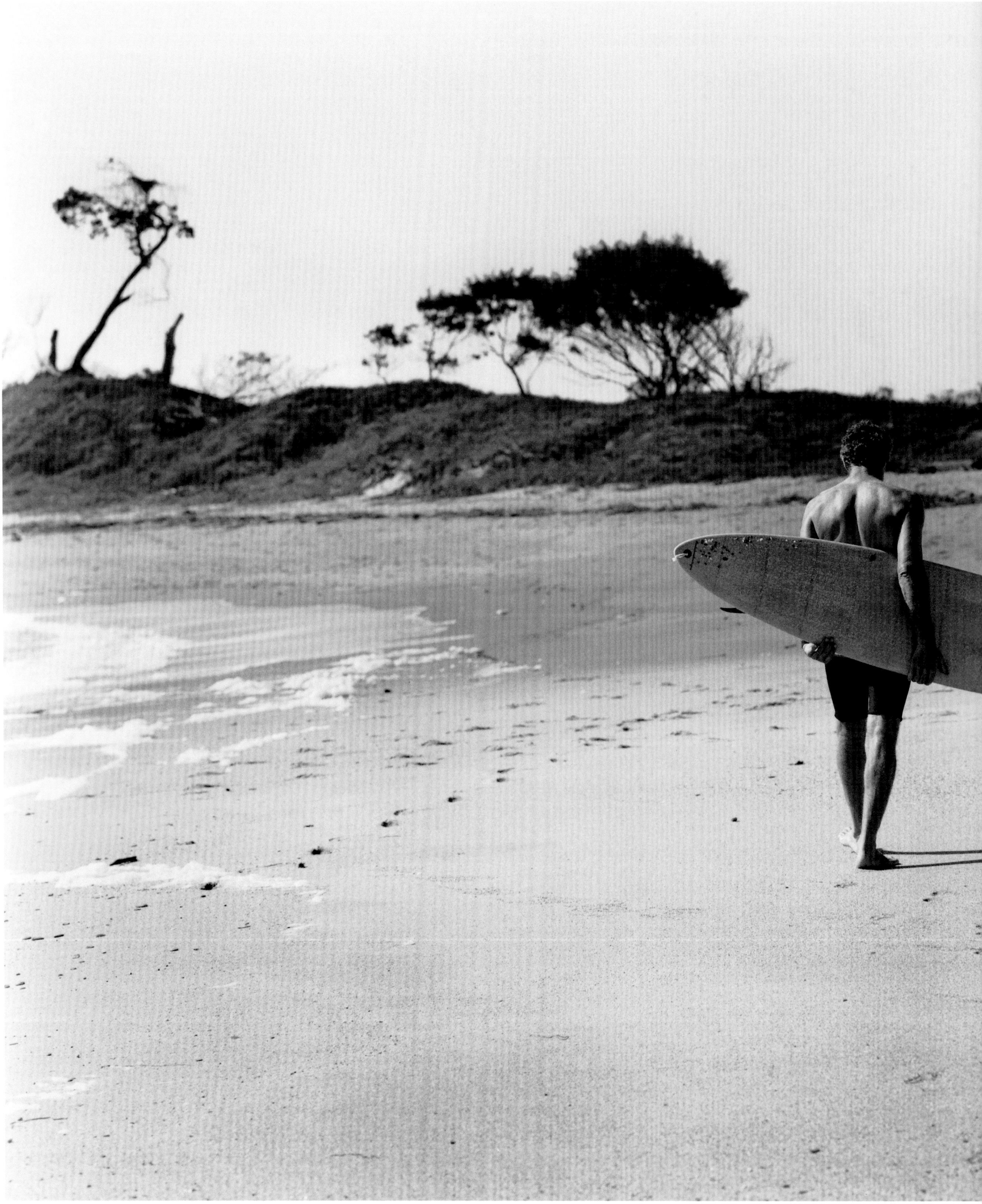

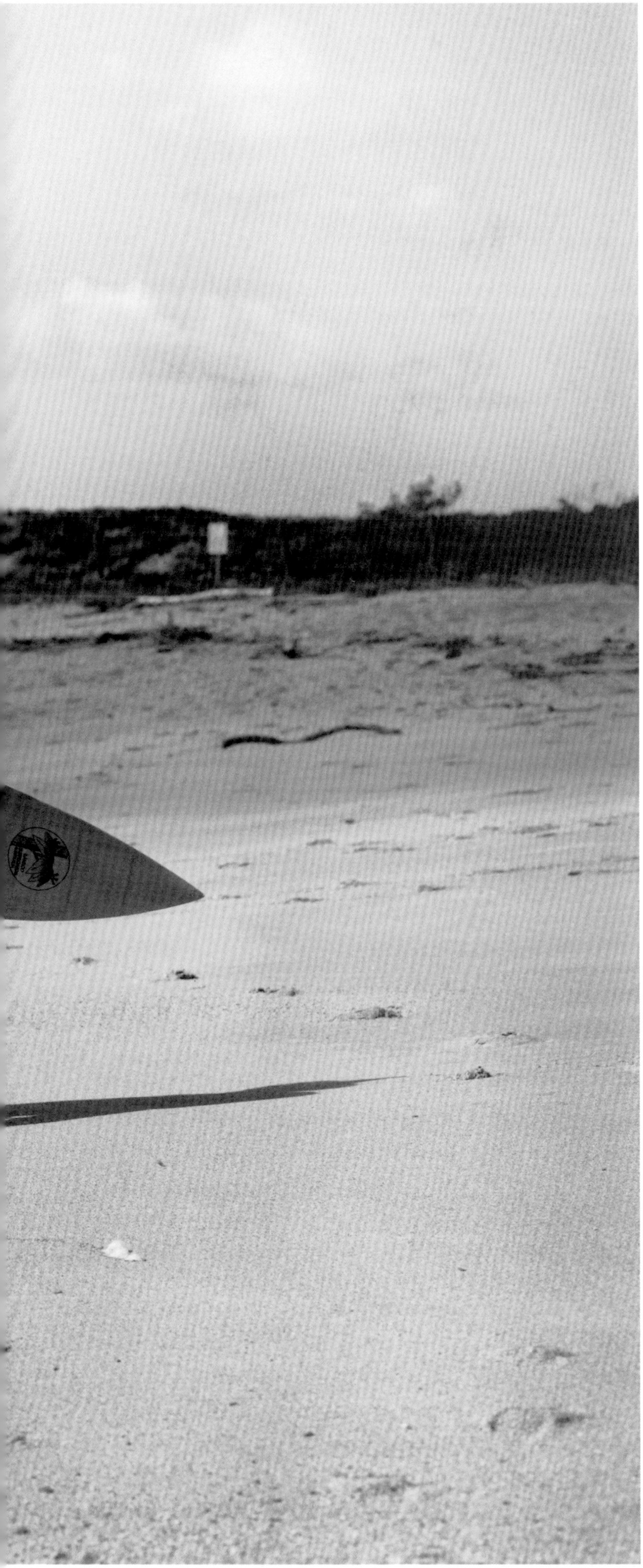

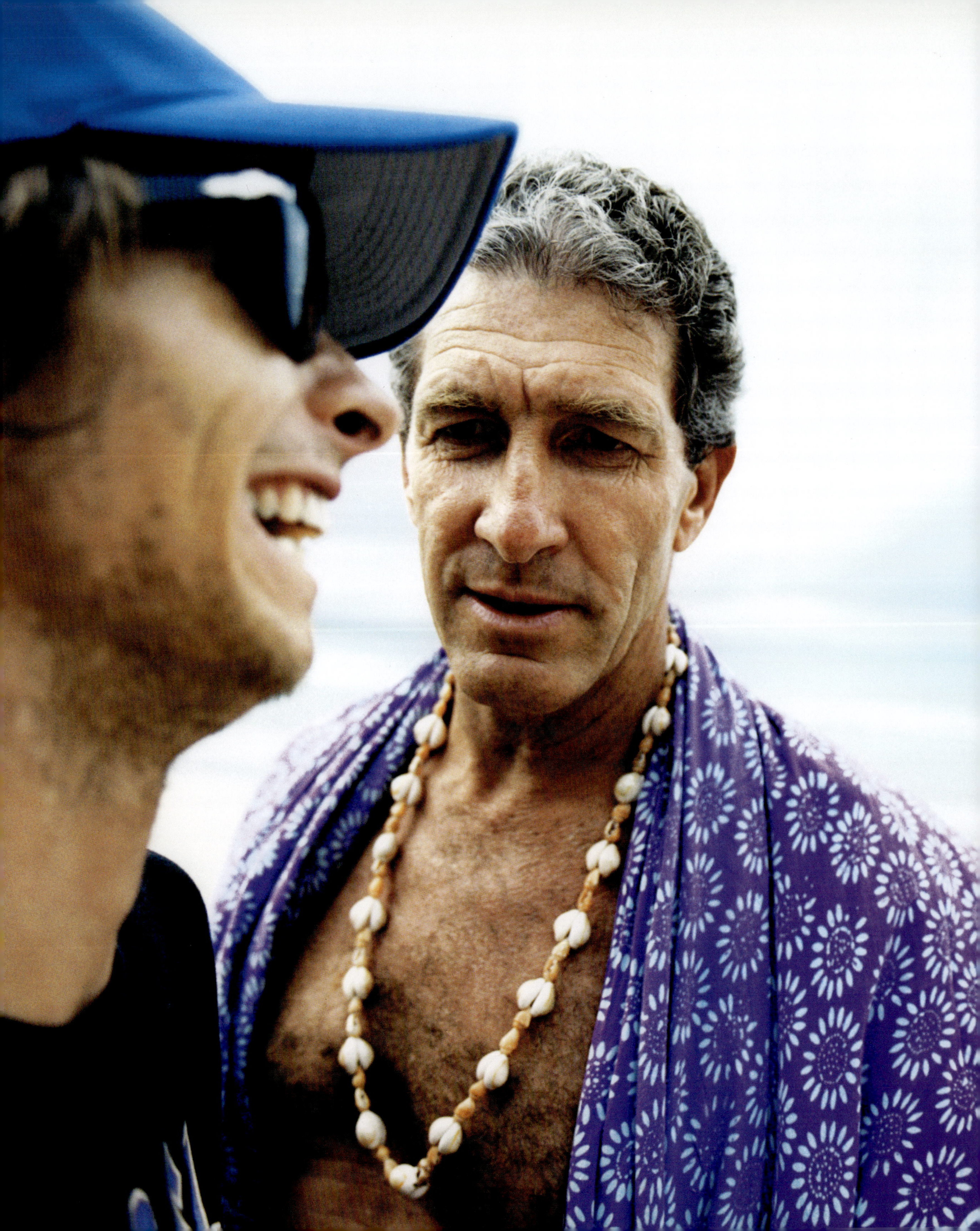

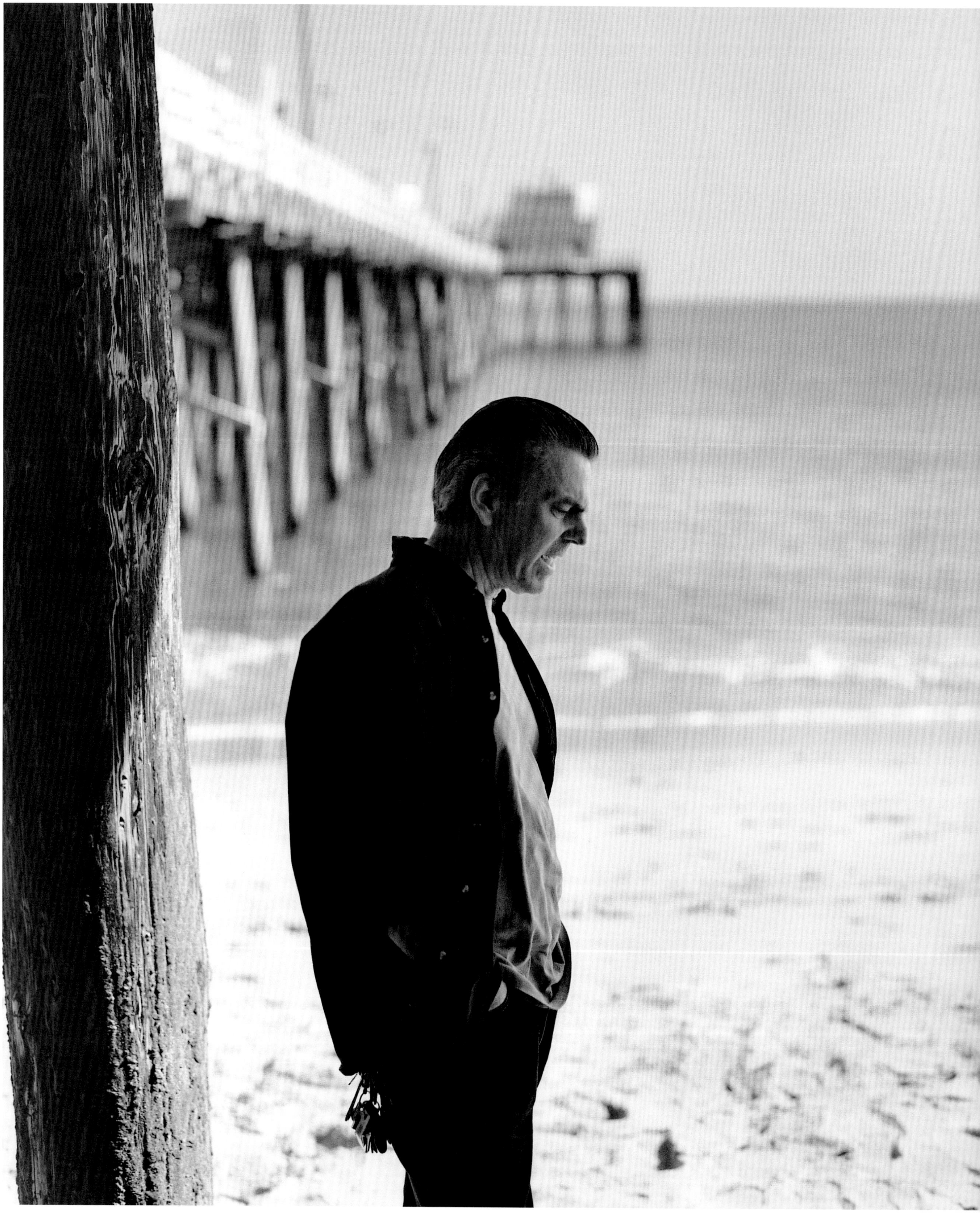

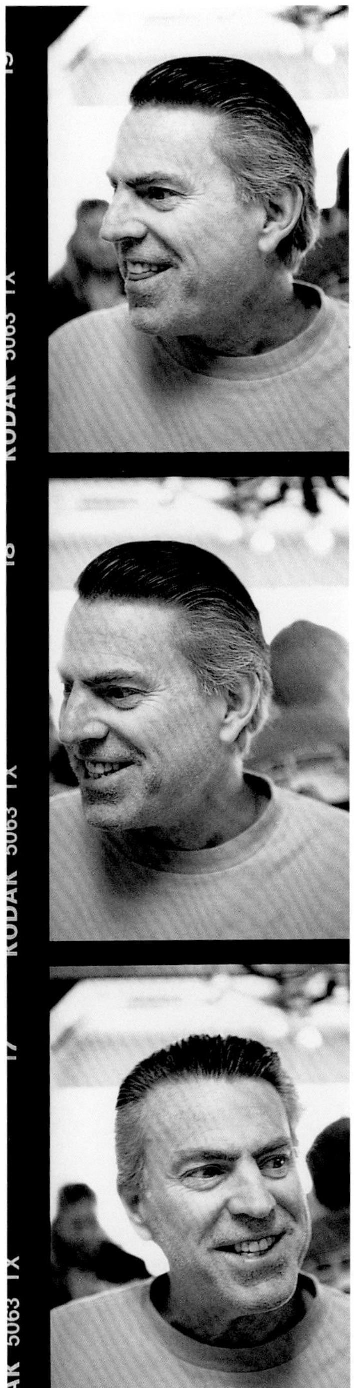

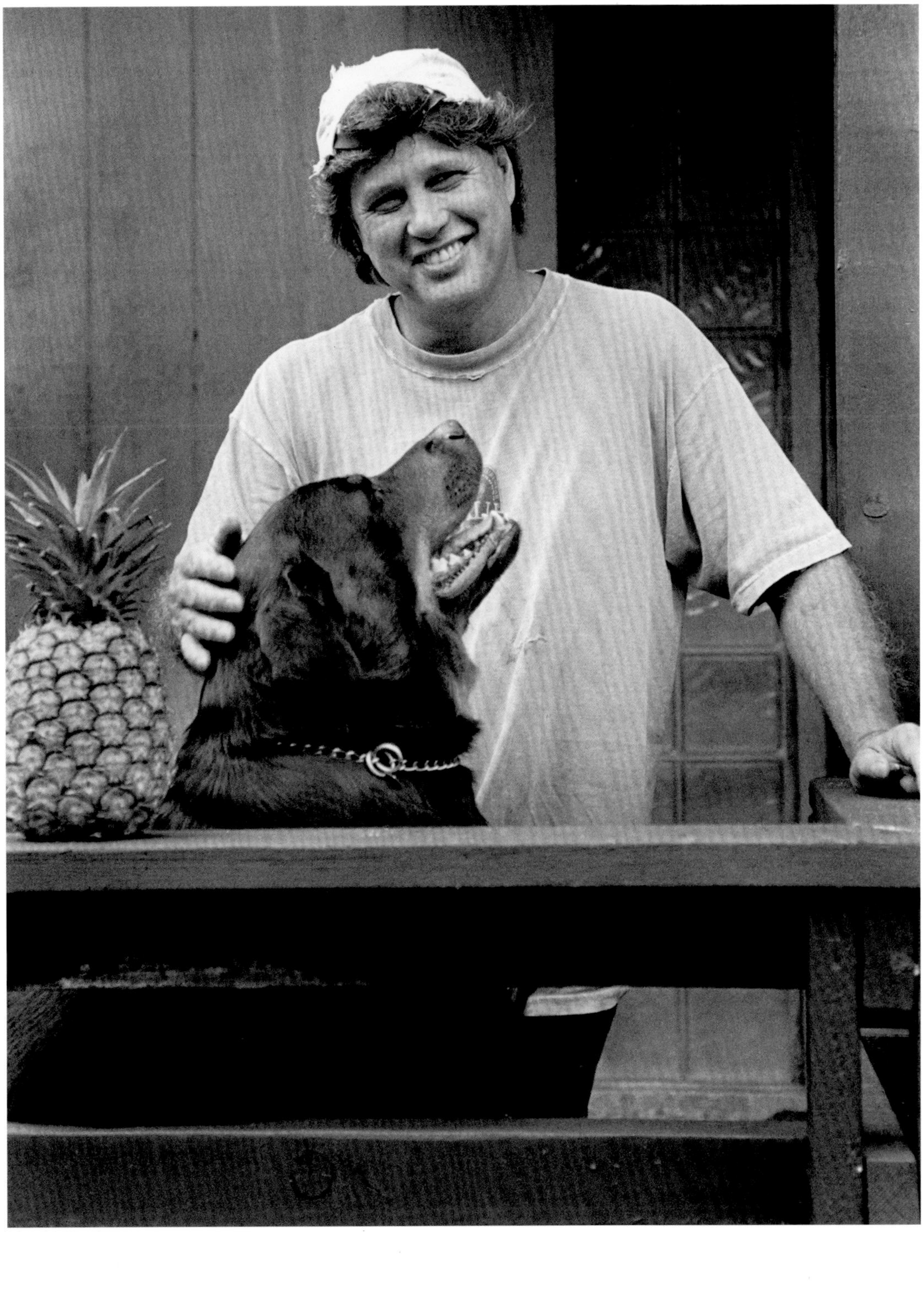

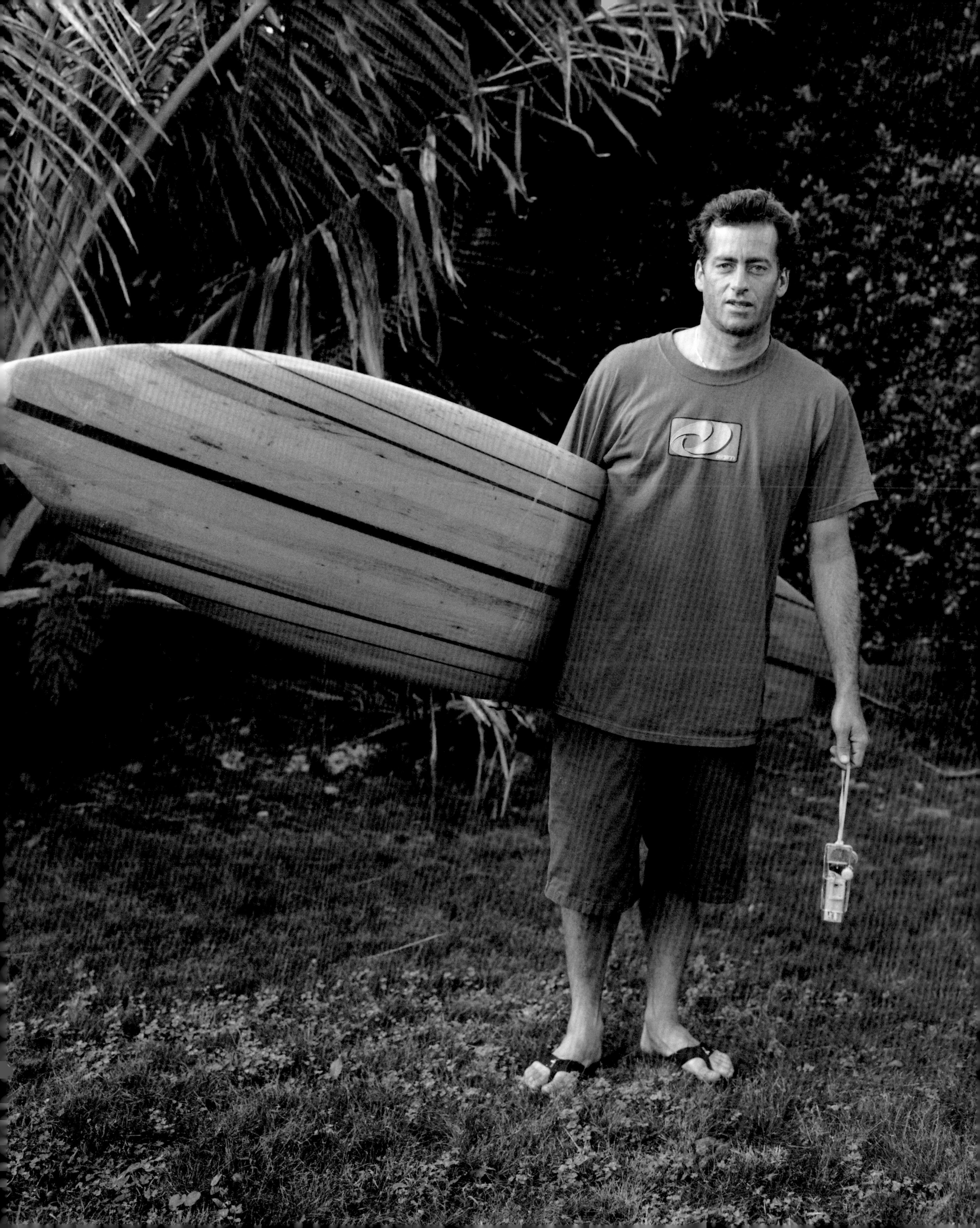

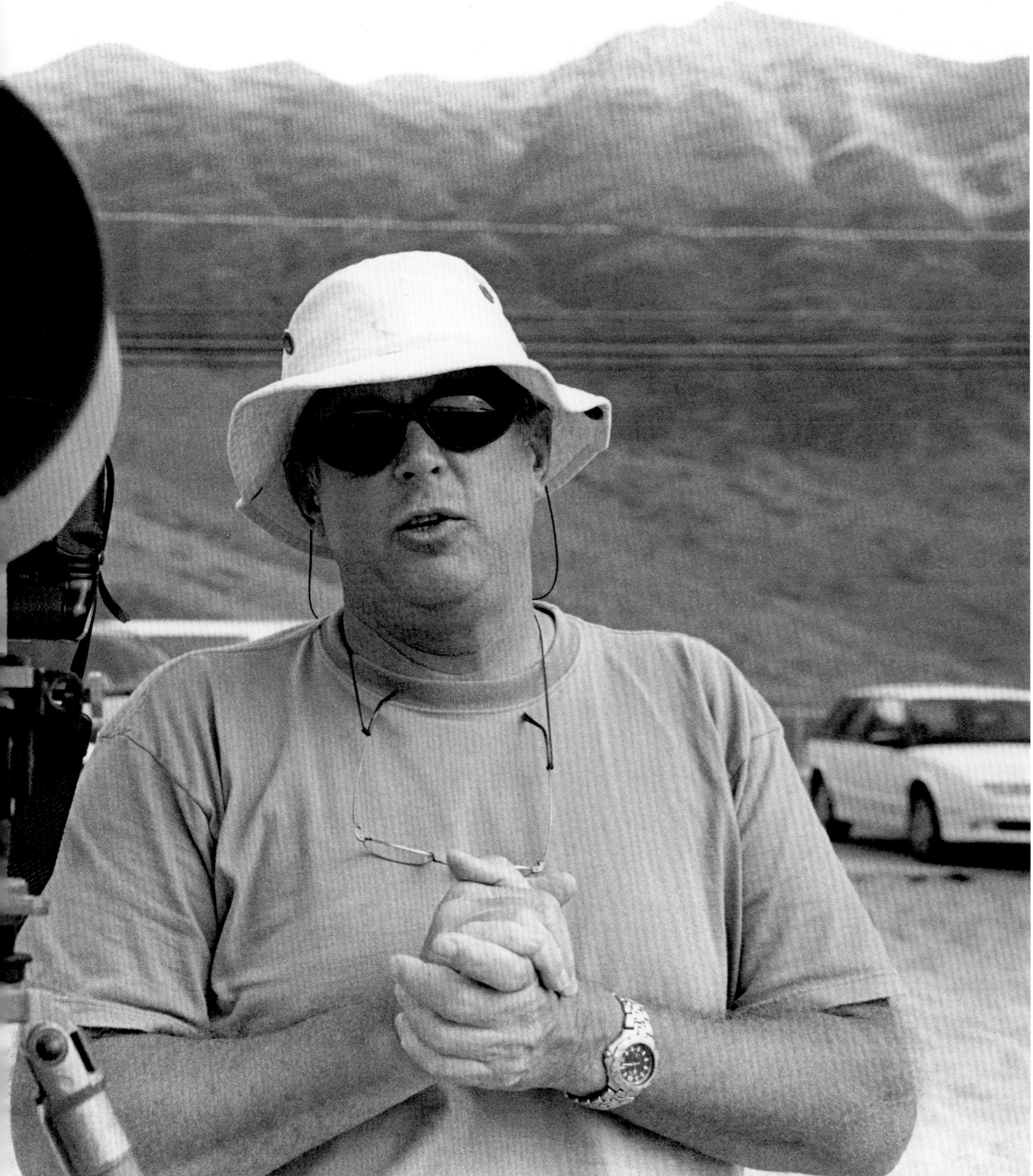

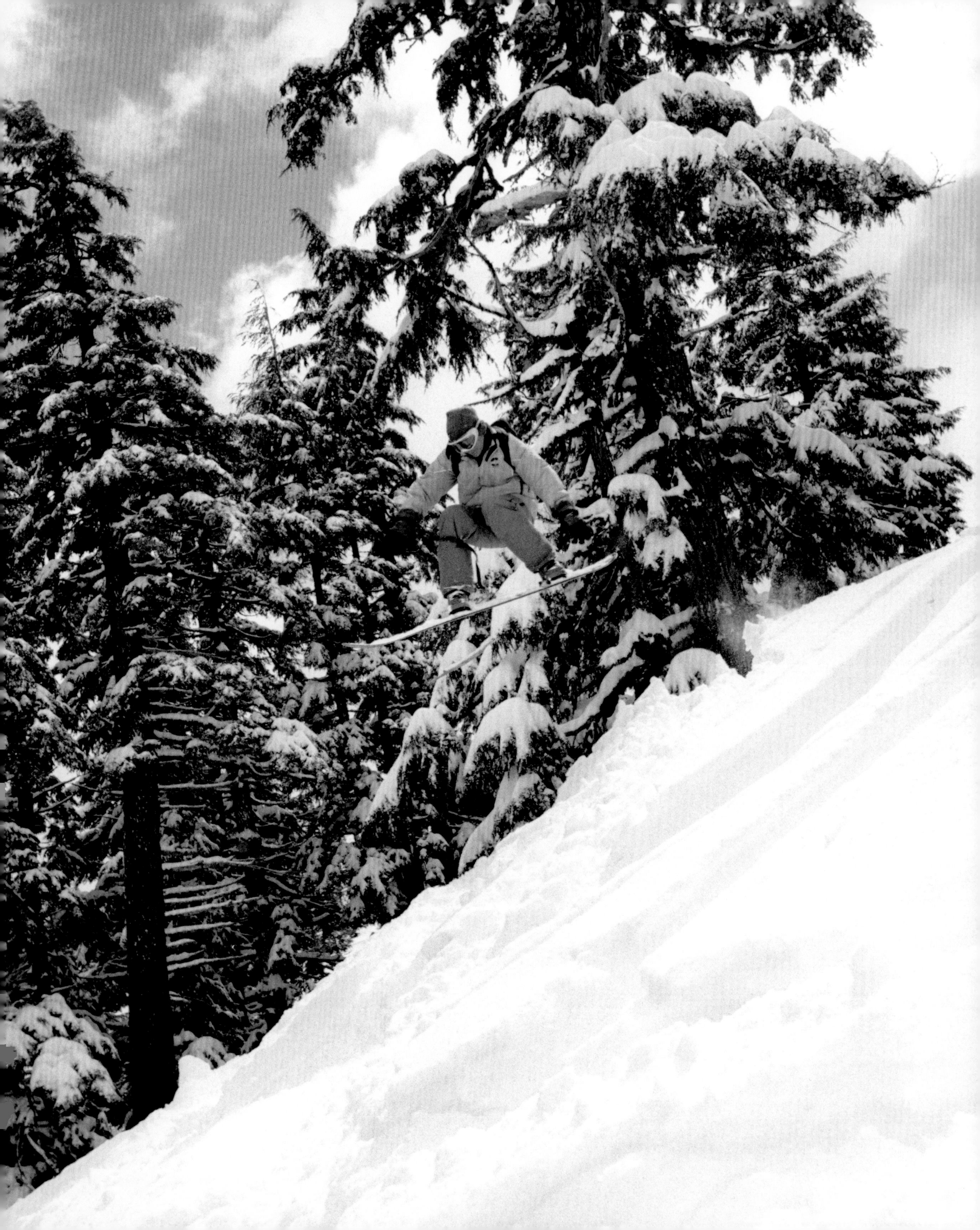

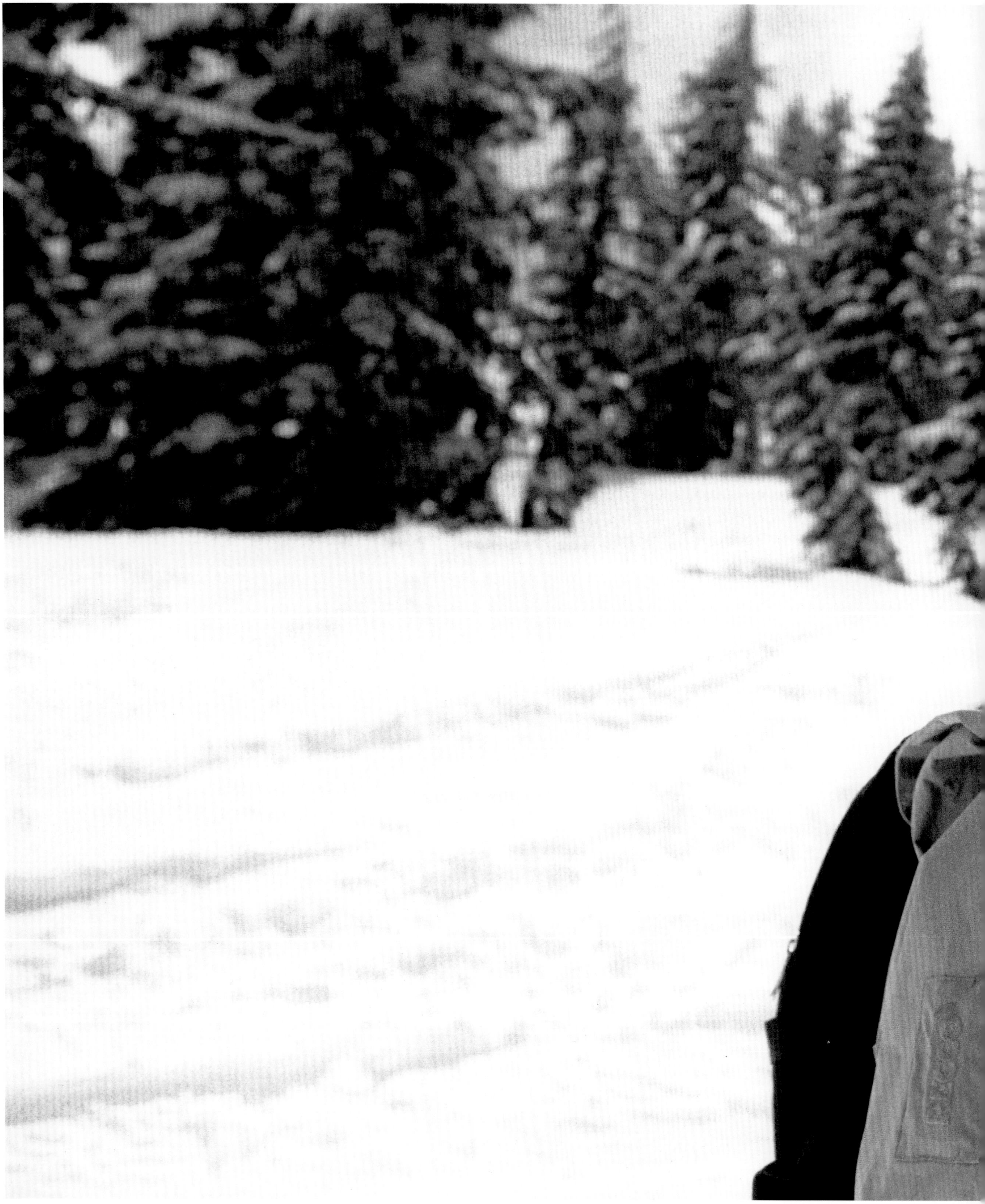

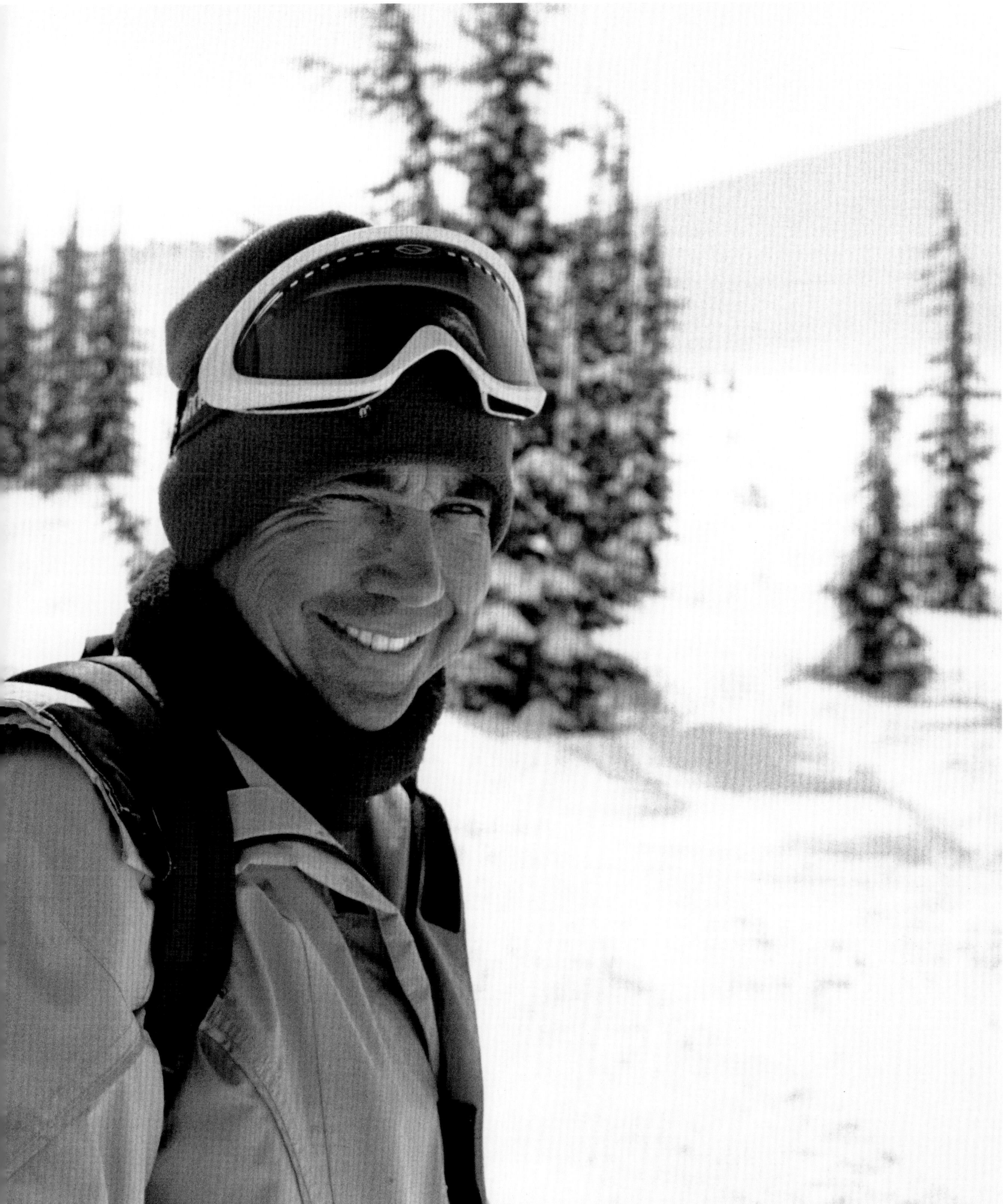

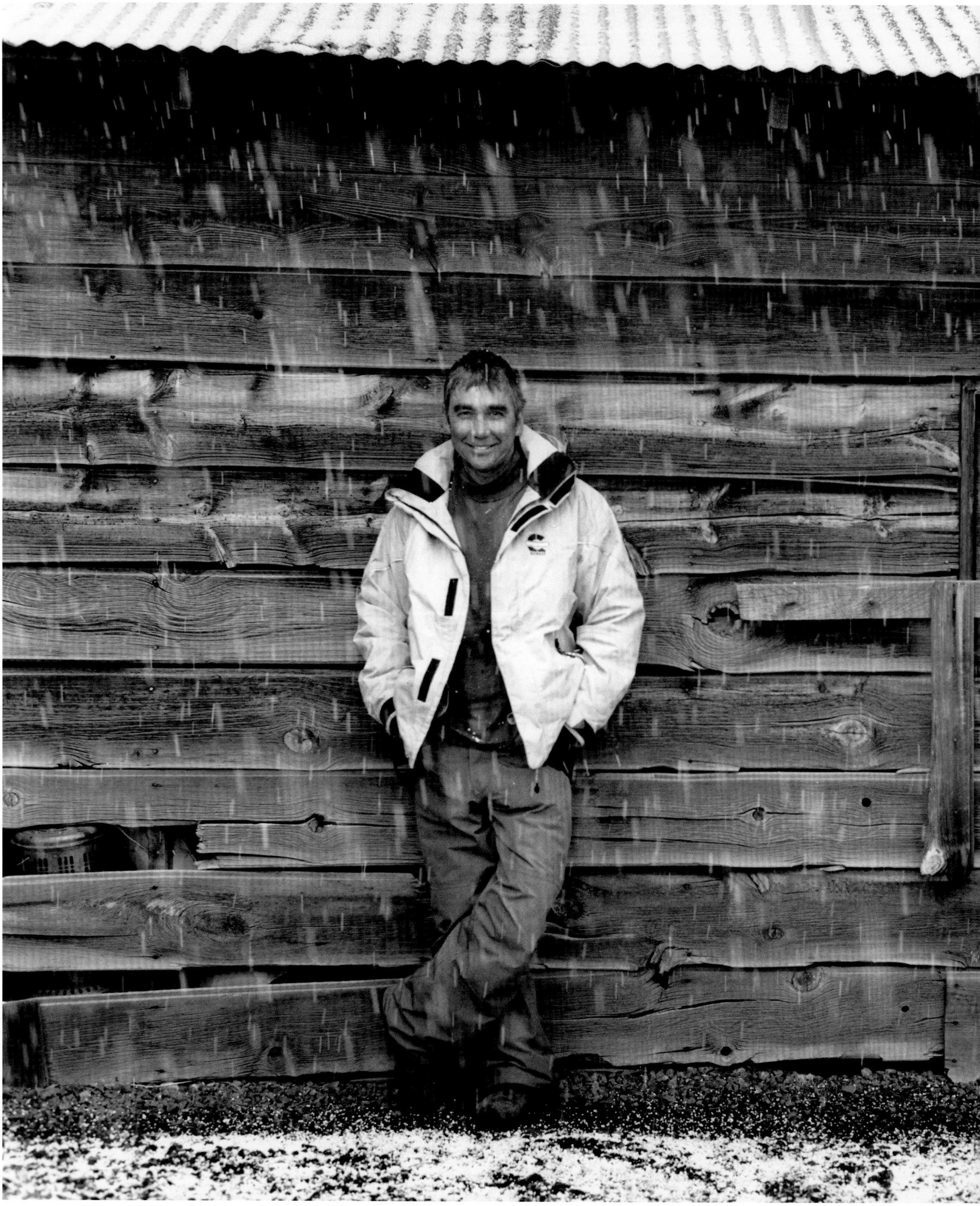

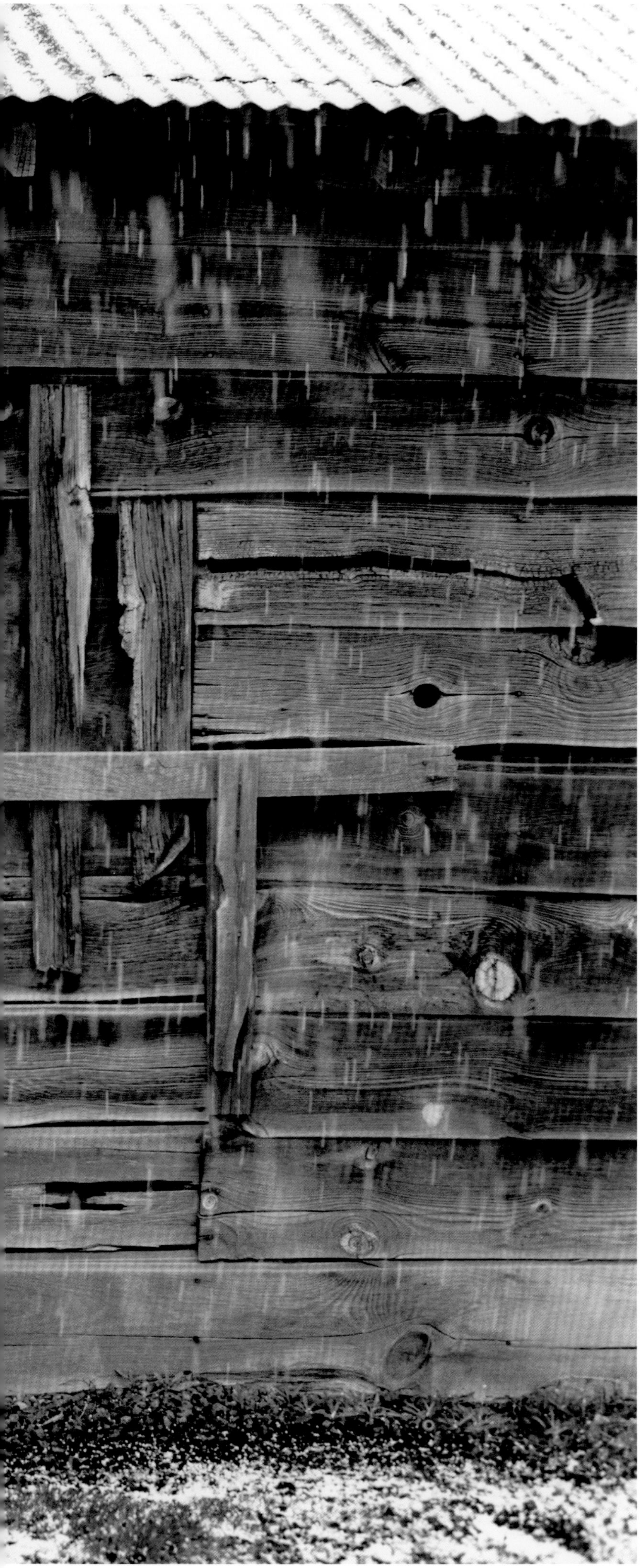

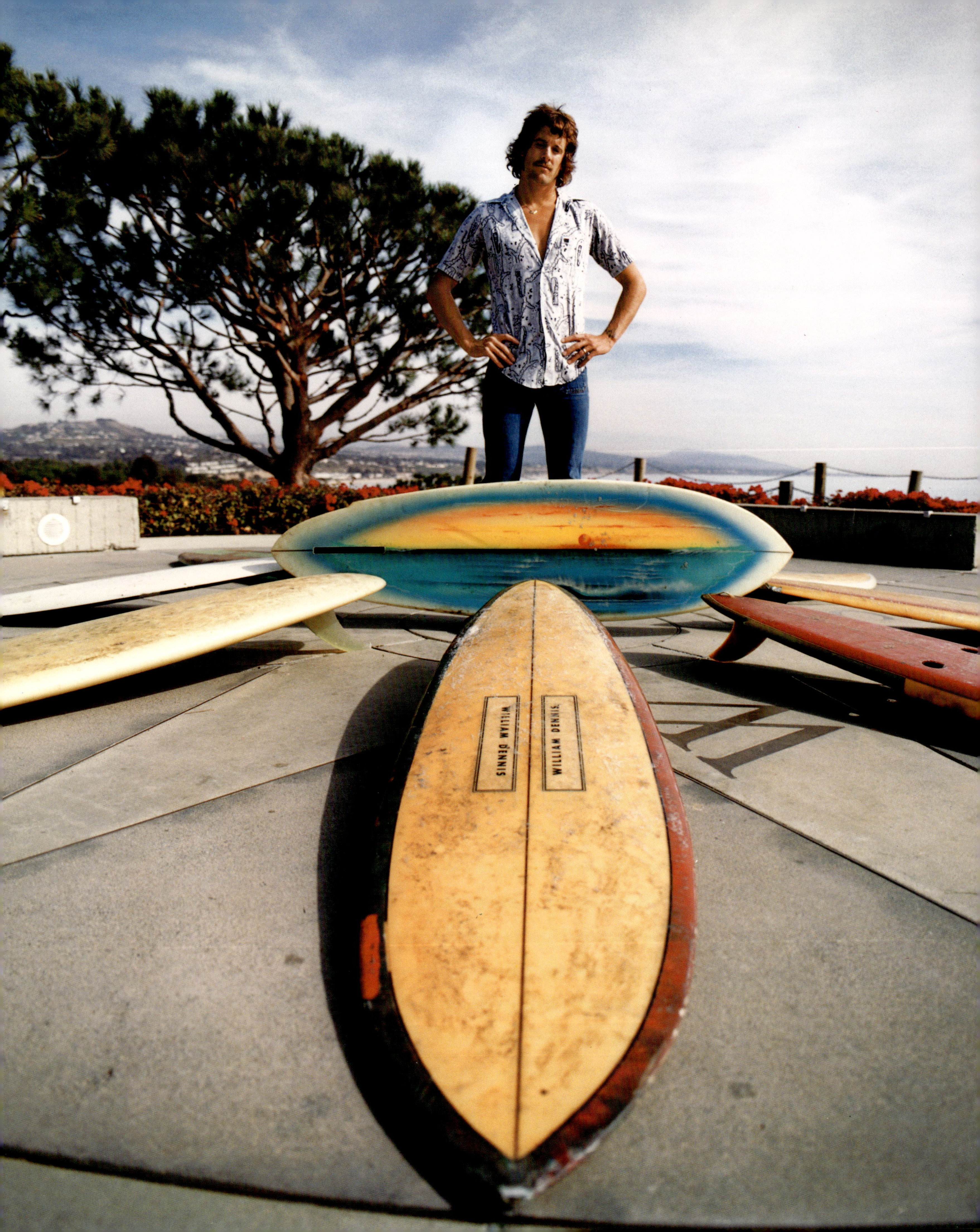
WILLIAM DENNIS
WILLIAM DENNIS

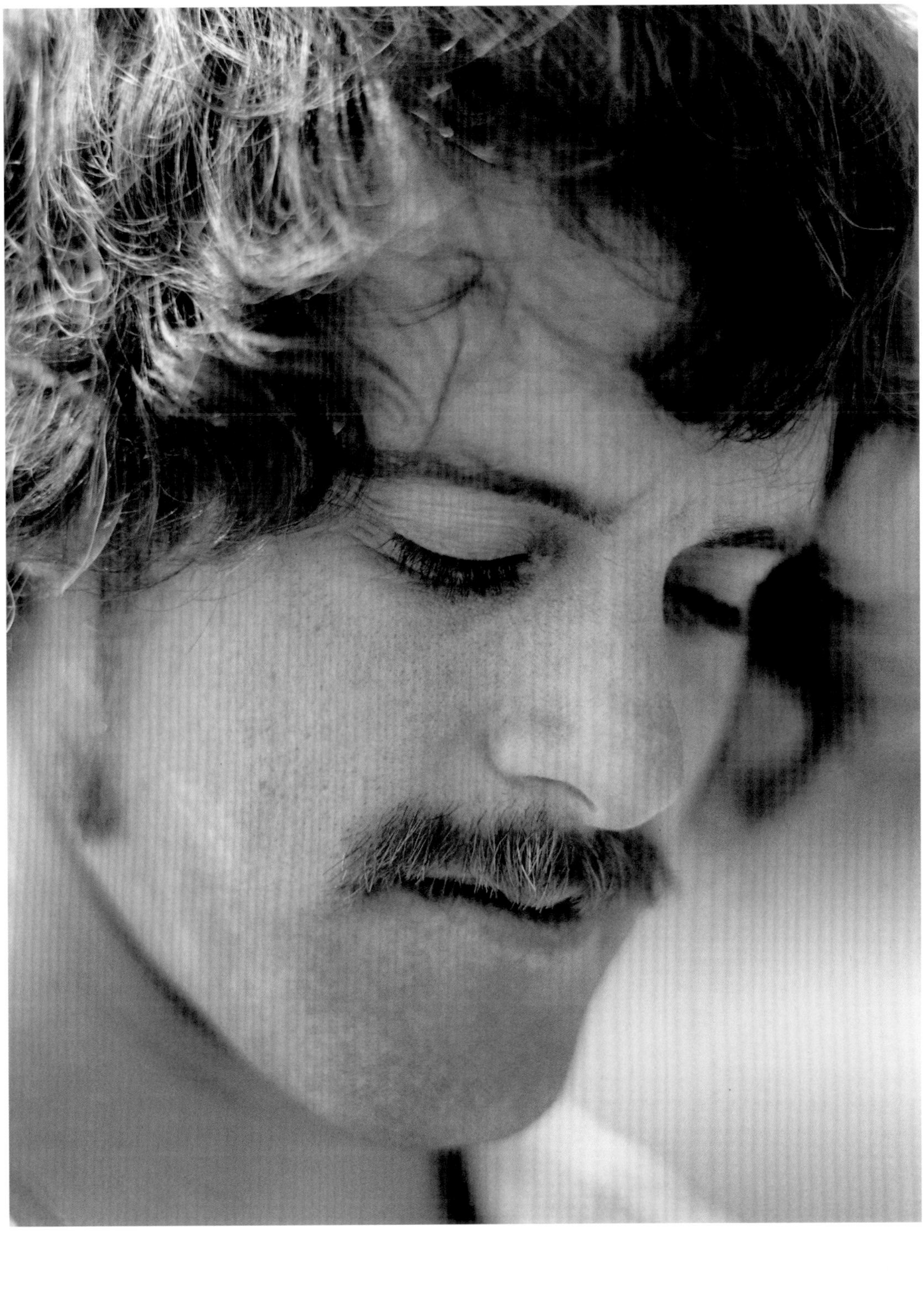

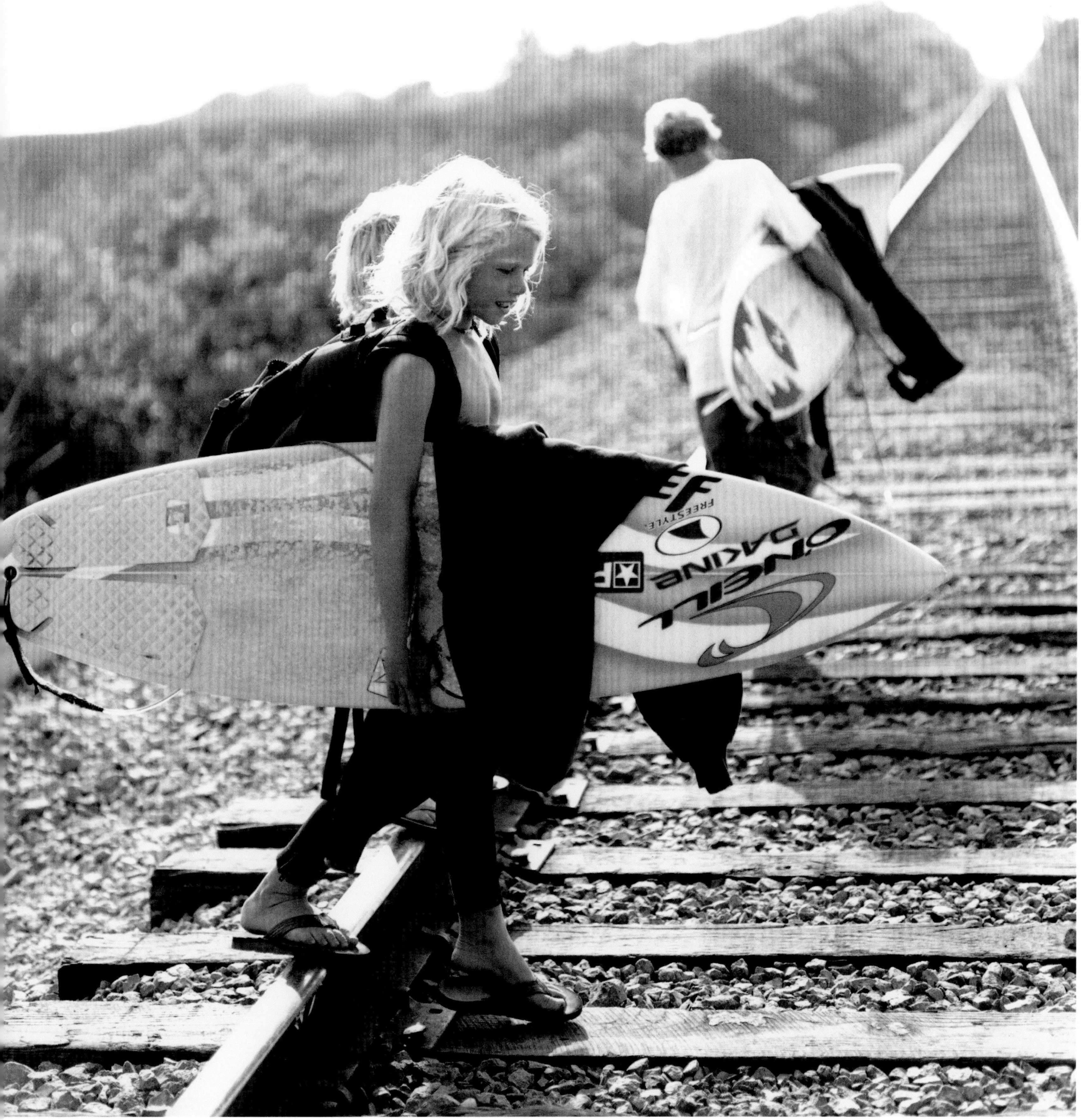
O'NEILL
DAKINE
FREESTYLE

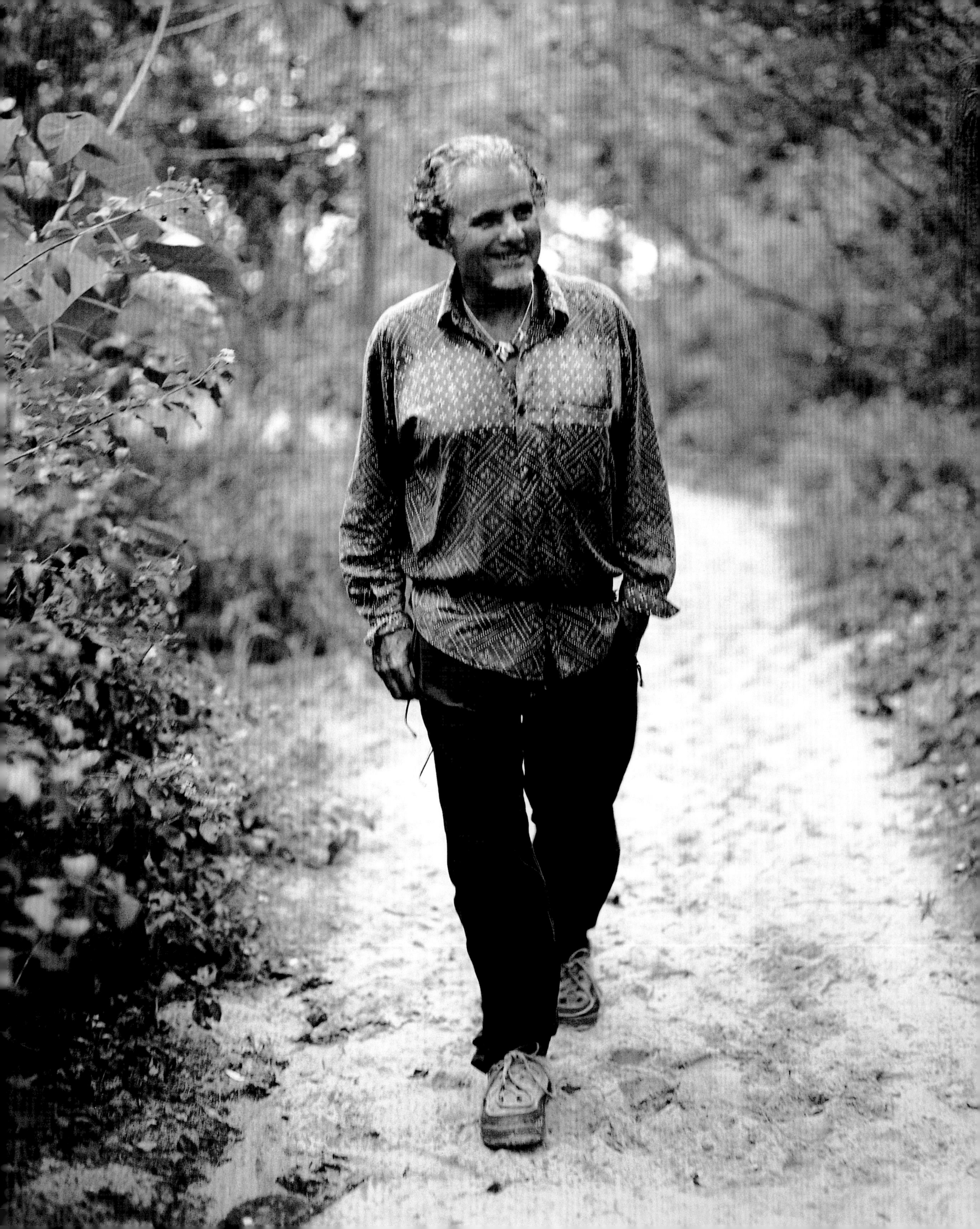

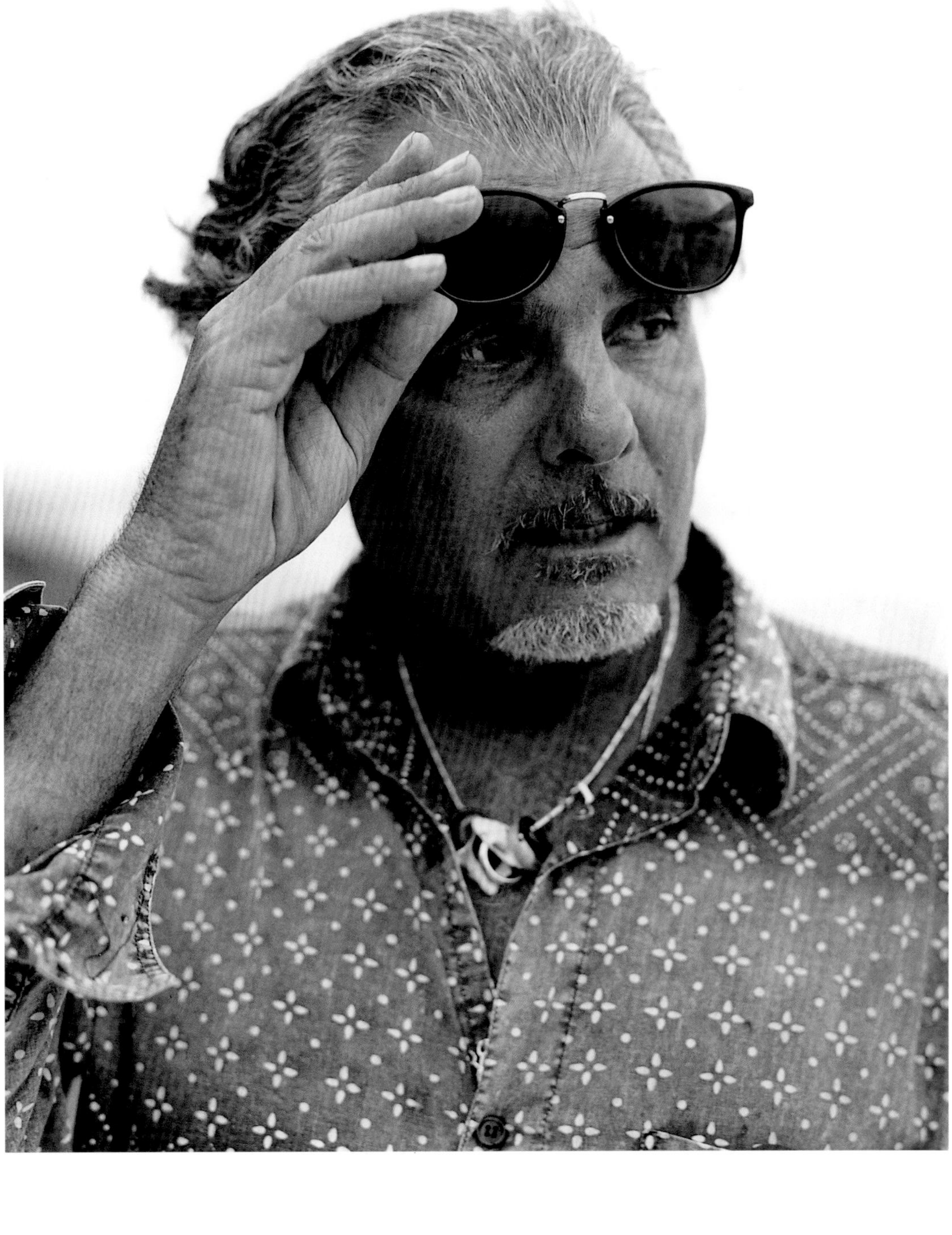

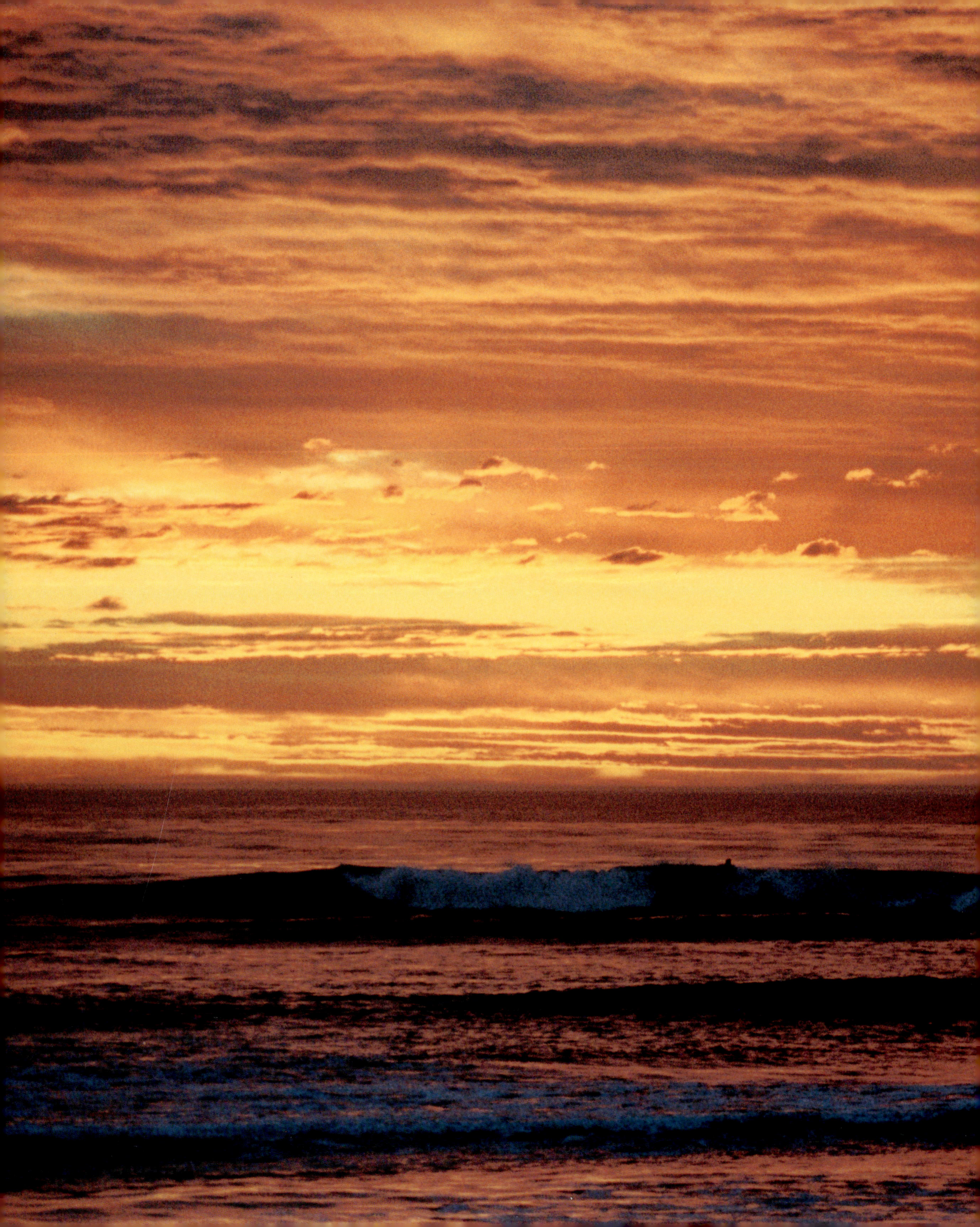

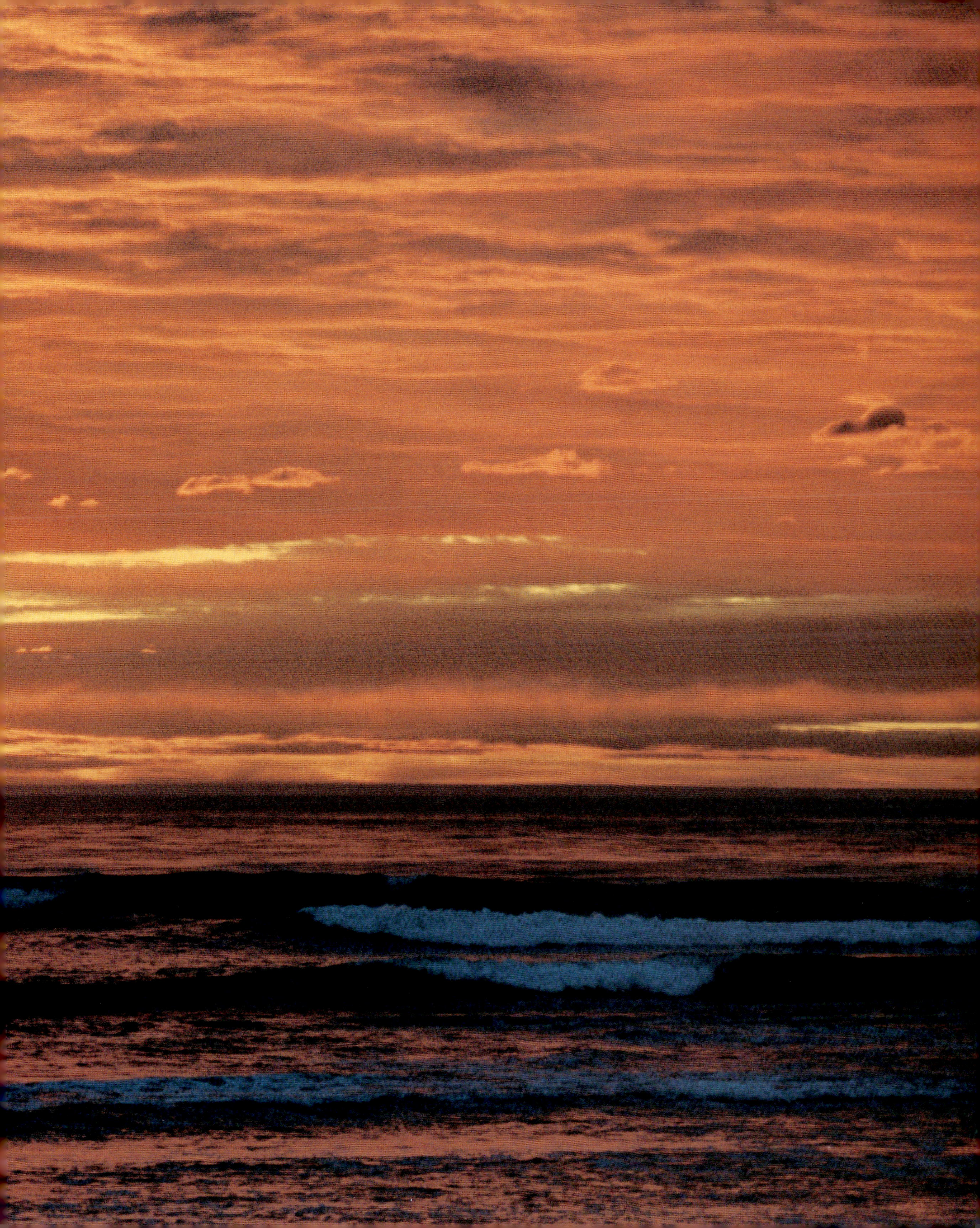

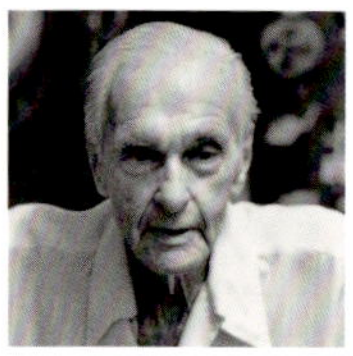

Bud Browne

I think a lot of younger surfers mistake Bruce and Bud Browne, and don't realize that Bud was working 15 years prior to Bruce, documenting and making films. Bud was the guy that invented the basic format for the surf film—the comedy, the skits, taking early icons and making them visible—all of that. He created some of surfing's first stars. For what I'm into, he's heavy. He documented the best surfers of the longboard period when they were kids. He has by far the best footage of Phil Edwards. And John Peck. And Larry Bertlemann. He saw talent really early; he's got an eye for it. He still doesn't get the credit he deserves.

Not too many people were swimming out to Waimea with heavy water housings, so his reputation as a waterman has been there all along.

Walking into his house you get a feeling of what he's about. The entire early history of surfing is stuffed into boxes and hanging on hooks on the wall.

He's still on top of it too. Even now, at 90 years-old, if you send him a video he'll check it out. He's into it, and he likes to know what's going on out there.

At the Surfer Magazine Poll and video awards, they had the audacity to tell him to "keep it short" when he gave an acceptance speech. Can you imagine that? The guy who invented what they were there to honor? J Brother gave him a copy of *adrift*, and he actually watched it and had some criticisms. He called Joe at 7AM and said "Well, I looked at your film and I have some criticism for you. I don't think opening up with such small waves was a good idea. Me, I always like to open with a bang."

He's really cool, and he really cares, even enough to take some time with a couple of youngsters like us.

A lot of people have the misconception that Bud's a vegetarian based on his build and how long he's been around. But he told me, "No, no, never worked for me. I just have a strict diet and stay away from the junk food."

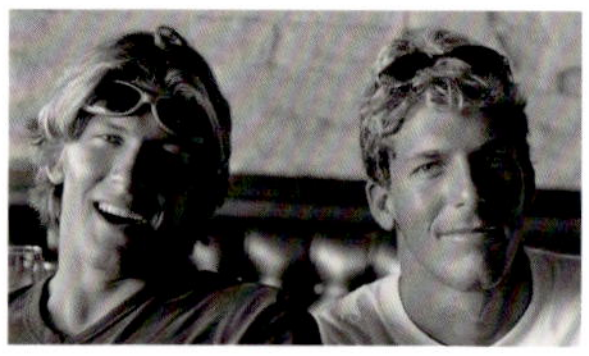

Bruce & Andy Irons

The Irons brothers are really grounded, and have a solid historical background. Their uncle Rick was a shaper and big wave surfer in the early '60s. Their uncle Jimmy is one of the most stylish people you'll find. If you surf Pinetrees, you'll see him playing his saxophone before he goes out for a surf. He surfs in Lacoste polo shirts, and holds his arms out when he rides, with his fingers just so, like he's pinching roaches.

That style trickled into the whole family, and it's really extra evident in Bruce with the flow he has going on. They both have an appreciation for things outside of high performance surfing. Growing up in Kauai they've been able to surf a variety of breaks, and to catch the influence of all of the character that's there, from the mysto-hippies to the Hanalei specialists. I've been out at Pinetrees and had Bruce come up and trade boards with me. He'd give me his 5'6" Merrick or whatever it was at the time, and he'd grab my longboard and do all sorts of funny poses, imitating his Uncle Jimmy... "Hey Joel, roach clips!"

They're the future of surfing. Kelly took it to one level, and they're raising the bar. At Pipe, Kelly's still in his domain, but they're the guys right now. Their sibling rivalry is unique in surfing, and I think it drives them.

Their security meets music industry standards. They're very well protected. They don't have to worry about some dickhead local trying to screw with them just because he's seen their pictures. They can go out anywhere and get waves. What's cool is that even though they have security, they're both gentlemen. You don't catch ego coming off them; they don't get caught up in the bullshit.

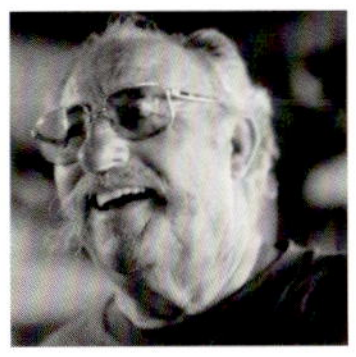

Dale Velzy

Here he is. The father of the surf industry. What's really unique about surfing is that our pioneers are still active and approachable. Paddleboarding, redwood to balsa to foam, Malibu, Hermosa, Dana Point, Velzy did it all. He saw the surfing scene bloom in the early '60s, and he built a commercial engine to run it.

The boards he's making now are like the Aston-Martins of the surfboard world. And they're priced to match! They should be. His curves are still as contemporary as any longboard shaper alive. Donald has one, and if you lay it on any good longboard, the lines match right up. He's the originator.

He's in his seventies, and there's not a bitter bone in his body. So many others grouse about the crowds and the way it is now. Not Dale. He's always treated me like a grandfather. "Hey kid, get over here. Check this board out." He pioneered Indo when he was in the Merchant Marine and Hawaii too, all back during WWII. Most people don't know that, but he'll tell you about it if you're bold enough to ask.

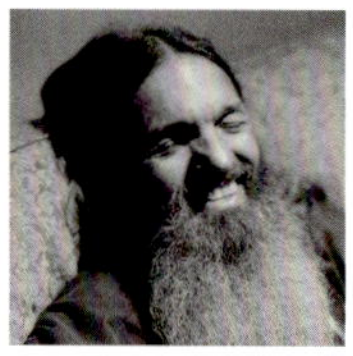

John Peck

He's in his 50s and he's still getting barreled. He surfs every day, does his yoga, and lives extremely minimal. I suppose he could live off air if he had to. I traveled to Australia with him for two weeks, and I took him to the Pass and Snapper Rocks. He was getting cheered into waves by people who didn't know who he was. They just instantly felt him.

We all know his background in the '60s, but he keeps an eye on us kids. He keeps us from going astray. Sometimes some of us go a little overboard, and to have someone who went through it all means something. You don't have to destroy your body and brain cells. If you're going to open a door, open it, but don't lose your mind.

He's a master of yoga, stretching, and breathing. Most of my dad's friends can't even touch their toes, but Peck is just an instrument.

His appearances at spots up and down the coast are becoming legendary, and he's noteworthy for more than his appearance and history. He rips. You'll hear about him all over. "Peck was at Scorpion Bay; he got the wave of the day." "Peck got so barreled at Swamis last week!" That's far out! He's in his late 50s!

He doesn't cut his hair 'cause it'll damage his nerve endings and disturb his energy flow. I can respect that. Something's working!

We get along really well. I met John at Donald's when I was 12, and he was recovering from his years of drug abuse, and he was scary. But he just kept working at getting it back...and he did.

One year at the Op Pro he wanted to sit in the spectator box, but they wouldn't let him in. He was just too freaky looking. I ran down and told them he was my uncle. From then on, we were friends. He gives us all adjustments and massage work. From his decades of hanging with body workers and yogis, he absorbed a lot of knowledge.

There's no apparent age difference between John and us; he just fits in seamlessly.

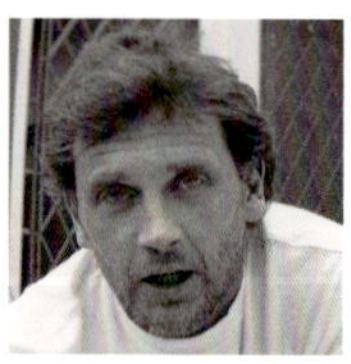

Wayne Lynch

Undoubtedly the best of the transition era, and I wish I could have seen him ride a longboard. From the photos, he was just as good on a log. I emulate his surfing more than anyone else's. There's something raw and pure at the same time in his surfing, and a hundred years from now when everything else is gone, his surfing will be remembered. He was a huge progressive influence; he changed the evolution of surfing, particularly on his backhand. He had ridiculous power, but he had such control that he rarely spun out.

I watched Evolution with him, and he just writes it all off. But I saw him on an egg at WindanSea and I'll never forget it. All the tough guys and derelicts in the lot were filled with respect for him.

I went to Tavarua with him and a few young guys. I was paddling out with Chris Ward, and Wayne got a tube ride in front of us. He was almost noseriding through it, going so fast. And to look over at Ward and to see his expression was all-time. "Did you just see that? How old is that guy?"

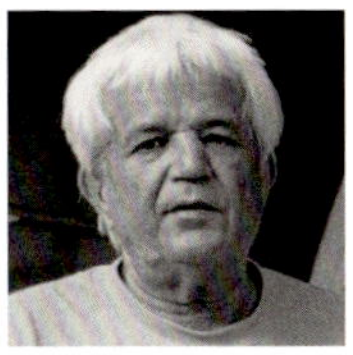

Dick Brewer

We can all say we owe some thanks to Dick Brewer. Any shaper who claims not to have some Brewer curves in his template quiver is full of shit. That's how influential RB is. We're talking a major contribution. Like Velzy with longboards, you can drop an RB template on any top-level shortboard, and the lines match up.

When you order a Brewer, you're not asking him to shape you a board from your head. You're there to hang out with RB and to take what he gives. It's a lesson in the shaping craft when you're in the room with him. From the moment he switches on the planer, he's talking you through it: "OK, I'm putting this vee in the bottom because this is what Laird's using at Teahupoo, but you're going to be at Pipe, so we're going to alter it slightly..."

By the time you leave, you feel like an astronaut. Some shapers drop knowledge while they're shaping, but in your head you're thinking, "Who the hell are you telling me all this stuff? Who has ridden your boards and did they work?" Whereas with RB, it's "I made this for Lopez." or "Reno won the Smirnoff on this."

I've had three opportunities to make boards with Dick. The first two were hit and miss, but the third was flat-out the best surfboard I've ever ridden in my life. It's on racks at home; I'm afraid to ride it. I don't want to break it. It's an egg. I wanted a Bing Lotus, and he couldn't figure out the template, so we talked it out based on his memory and advertisement pictures I'd seen. When we were done, he was afraid to sign it; he wasn't feeling it.

If you ever have the chance to get a custom Brewer, keep your ears open because he's a master professor.

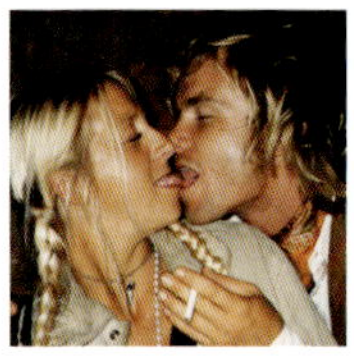

Matt Howard & Brittany Quinn

There's so much talent between these two that it's sickening. The longboard trip is filled with colorful characters. Between Britt's sewing and Matt's animation and their surfing, their filmmaking, the way they carry themselves...

Matt and Britt are two of California's true gems. There's not a surf couple like them; they're inseparable. They gave a true identity to hetero surfing. They have so much style and mystique, and none of it is fake. They're UCLA graduates, and they're totally tweaked and real. They built a life around point waves, longboarding, and art. That's flavor you don't find elsewhere in the modern surf world. You won't see that in the magazines.

They're just trying to survive, to hang in until Rincon or Scorpion or Malibu breaks. Thank God for vagabonds.

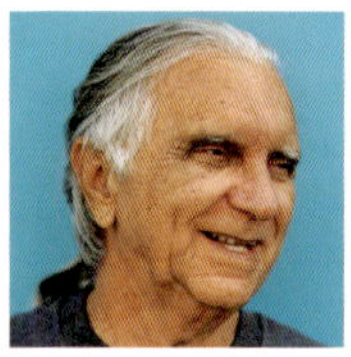

George Downing

I'm in awe of Mr. Downing, from his paddling, to his big wave surfing, to his design work, to being the guy who makes the call on the Eddie Aikau contest. It's all about respect. He's the godfather. All my life, I heard "Downing's the guy." You go into his shop, and it's filled with all of those beautiful wide-point-forward guns. He made surfboards work in huge waves. Nat Young clued me in to George. They don't get along, but Nat made it clear to me. "Joel," he said, "That's George Downing and you'd fucking better know who he is and what he's done." It was obvious that Nat holds him in high esteem.

The more homework I do in board design, the more I learn that he's at the top of the family tree.

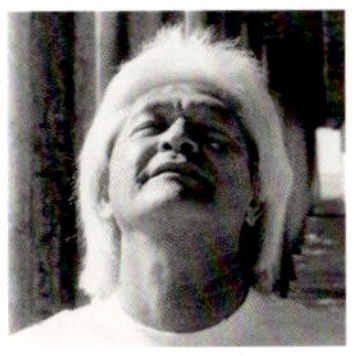

David Nuuhiwa

Up untill 1966, Nuuhiwa absolutely captured the peak of long board surfing. He was the embodiment of the pinnacle of surfing in that period. Even compared to the noseriding that's going on today—and I think it's at the highest point it's ever been—if you look back at the old film footage, David's still untouchable. he was originating all of that stuff. Everything we do now, came from him—the weighting, the lifting of feet, he was the man. Totally inspirational. In the shortboard world today, no one looks to guys from earlier eras for inspiration. But if you ask a 12 year-old longboard kid who Nuuhiwa is, they can tell you all about him. We're all just emulating his grace on the front half, and he'll never be forgotten.

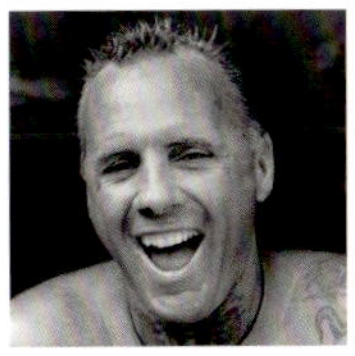

Jay Adams

I didn't know about Adams' skate heritage when I met him.

I was a teenager surfing Velzyland, and I was getting super hassled by the local kids; all but ordered out of the water. All of the sudden this tattooed haole guy turned around and calmed them all down and told them to be mellow. That was Jay. He'd seen the Santa Monica Airlines sticker on my board, and figured that I must be hooked up back in his hometown.

Jay's dad was a classic Malibu surfer, so Jay has an affinity for surfing outside of the mainstream...and that would include longboarding.

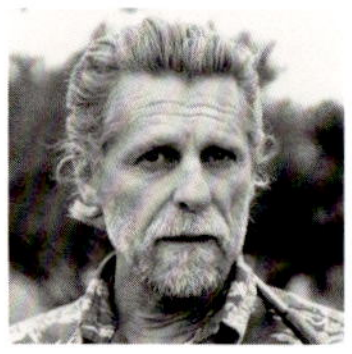

Michael Hynson

Hynson nailed it. He gave surfing its identity, as far as I'm concerned. His look in *Endless Summer* was untouchable, with the Ray-Bans, the white WindanSea t-shirt, the trunks, the slicked back hair, the transistor radio...he was so cool he was too cool. He took the Hispanic vato style and tweaked it to the surf side. He defined the surf style. Maybe not so much to the surfers already doing it, but to mainstream America, he is the unforgettable face of surfing.

People write him off now, "Oh, he got involved with drugs." Who didn't in that generation? It's a shame to me that he never benefited from his definition of cool in *Endless Summer*. He wasn't even allowed to participate in *Endless Summer II*. They have these *Endless Summer* reunions, and he's not even invited. How's that for audacity? Grown men holding these petty grudges just astounds me.

At a trade show, I was walking around with Hynson Jr., and Gerry came up to us. He told Jr., "I want you to know that your father is one of the reasons that we were able to ride Pipeline. It was a board that he made me that was the first downrailer that worked." Gerry knows. That meant a lot to Jr.

Reno Abellira told me that he owed money to Gerry for rent, and he gave him his 7'10" Hynson instead. That's the board that opened up the Pipeline for Lopez.

Hynson made state-of-the-art, foiled-out surfboards. The Red Fins were the joint. He's still a master. He shapes in slacks, leather shoes, and a nice dress shirt. And he's tough. If you even think of bullshitting him, he reads it instantly and he's over you.

His contributions are many. He put the cool in surfing, he opened up previously unrideable places, and he had a beautiful, Phil-inspired surfing style.

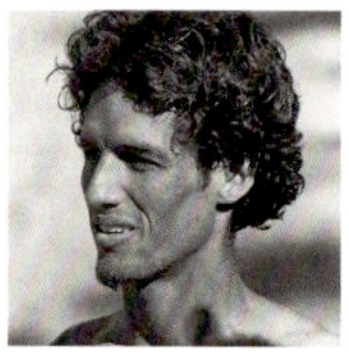

Rob Machado

When I was growing up, Rob was the only guy I would seek out to watch surf. He probably doesn't remember this, but when I was a kid I had my parents take me to a contest in Del Mar just to watch him surf. The moments when he'd come surf Cardiff were amazing, because you'd just say, "God, that kid has crazy flow." Fluid and so fast. Never wanky. He was the only reason to watch those *Momentum* videos, style wise.

Rob has a unique and beautiful style and he pulled it off on a professional level. He's the only surfer on the WCT with that much style. Rob can log, too. It's obvious that he could have an insane skill level on a longboard. He has an appreciation for guys like Joey Cabell, which speaks volumes.

Now that he doesn't have to compete, he's entered a new realm, and he's doing some interesting work with surfboards. We'll hear a lot from him in the future.

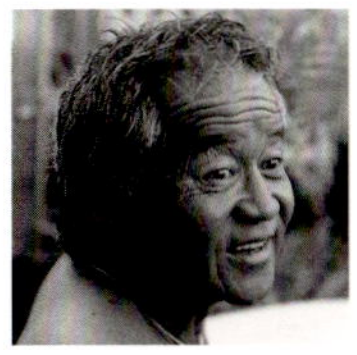

Donald Takayama

The most influential person in my surfing life. Donald teaches by example. He gave me everything. He spotted talent in me when no one else would give me jack. The opportunities he presented went way beyond the boards he made me. Let's just talk about the barbeques, the introductions to the pioneers of surfing. It took me until I was 18 to realize what he'd done. He introduced me to Miki Dora, Nat Young, the whole family tree. He gave me all the chances to be whatever I could be. He's a giver. I've literally seen him give a person the shirt off his back.

He's been mowing foam—hell, he's been mowing wood—for nearly 50 years. I don't think there's anyone in his class, craft wise. Watching him shape a surfboard is one of the most painfully beautiful things you'll ever see. When he's through with a blank, he doesn't even need to sand it. It's perfect. His finish work is untouchable. His product is so tight, he never has to stress. Donald produces 100% quality. He oversees every stage of production, every single day. And I'm talking Christmas morning. He gave up everything when he was 14 to become a shaper! School, family, the works. Flew away from his little island to work for Velzy...at 14!

Who else has done that? Who can even carry his planer?

You can look down the rail of a Takayama and know it's his instantly. And he can ride! He never reached the height of competitive success because of all of his other interests...mostly hot rods and women, but he blends Hawaiian beach boy style with a Fred Astaire level of performance, right down to bowing after a good wave.

Donald orchestrated my entire existence. His generation amazes me, compared to how spoiled we are today. They didn't buy houses with what they made from surfing, yet it's invariably his generation that gives the most. He took valuable time from his career to form mine. That sums him up. As a result, he's the ultimate hanai uncle, and he's treated like royalty wherever he goes. During a time when I was getting so much attention, Donald didn't like that I was overshadowing my brother, so he taught him how to shape. You don't forget soul like that.

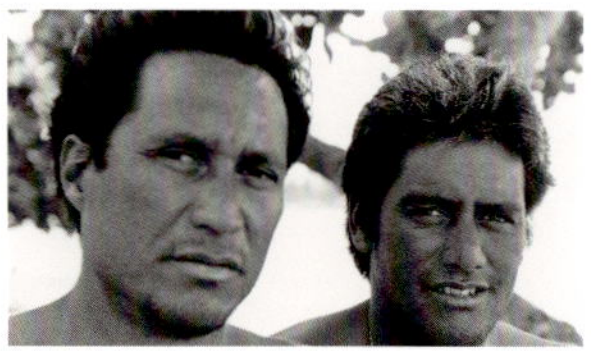

The Keaulanas

They live off the land. Not too many people have their own beach. There's so much beauty in their lifestyle. At 14 I went to Japan for the first time by myself, and I was a little tripped out. The guy who picked me up spoke no English, so he took me to Rusty and Buffalo. They looked out for me, and took care of me in Hawaii. Buffalo used to pick me up at the airport because I was too young to drive. I spent a long time at Makaha. I used to play Millipede on Nintendo with Buff. That family showed me a lot of love and they certainly taught me about respect.

They taught me valuable lessons, and let's leave it at that.

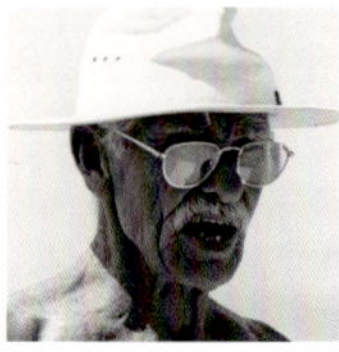

Dr. Dorian Paskowitz

The god of fertility. Any man who can father thirteen talented, attractive kids and home school them in a camper all over the world, surfing the whole time...well, that's heavy. The kids are all worldly and experienced. They're a living backlash to the status quo, as far as families go. Doc is a strong enough person that it worked.

When he was 75 or so (he's 83 now) he went to Tahiti with a tent and camped for three months. Didn't know anybody, he just needed to get away for a while.

And living in Hawaii before the war, he had it at it's best. He took the gift of surfing to Israel. And he's always lived a vegetarian lifestyle, and his body is a testament to that.

The thing I love about Doc the most is his brutal honesty. When I was competing against his son, he'd never mince words. Usually everyone bullshits you: "Ah you ripped; you got ripped off." Doc would say, "Now Joel, you did very well, but you lost. My son Israel just outclassed you today."

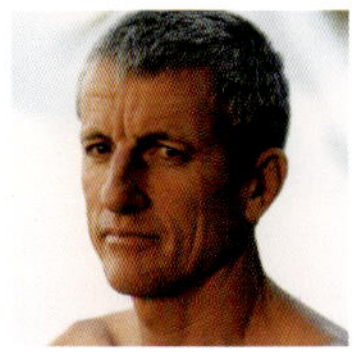

Mark Cunningham

My first encounter with Mark Cunningham was sitting on the couch at Gerry's house, and Pipeline was big, ugly, and really scary looking. It was one of those days that you just don't leave the couch. Then, looking up the beach, I see this guy with a Speedo, fins, and a bathing cap running down the beach. We were thinking, oh, this should be good; this guy's going to get killed. The first wave he got was the most incredible thing I've seen at Pipe.

He surfed Pipe without a board. I mean, he dropped in, bottom turned, got tubed, and then cut back.

Some contest organizers flew him to France. In Biarritz, people sat on the seawall drinking wine and watching him. They did double takes because he looked like a dolphin.

It was later that I found out that he was a superstar lifeguard with a rep for swimming out at Pipe and blowing everybody away.

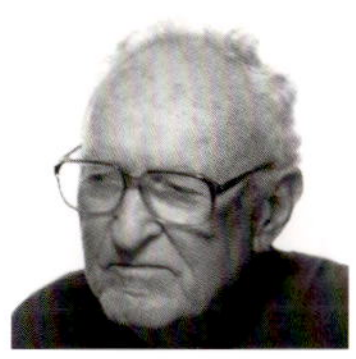

LeRoy Grannis

The longboard kids were fortunate in that the older photographers came down to the surf contests and sold prints, hung out, and talked story. I got to meet Don James, and of course LeRoy was always there. We got access that nobody else got. They'd sort of lost touch with contemporary surfing, but the longboard scene felt familiar to them. Grannis donated photos for the trophies, which really made them worth winning.

His photos were hugely educational for me, from a surfing point of view. The Dora photos, Nuuhiwa, Peck, all of them...his photos have a timeless vibe. Through him, I was able to get a good idea of what was going on in the 1960s.

LeRoy is one of the most crotchety guys you'd care to meet, but down deep he's a sweetheart.

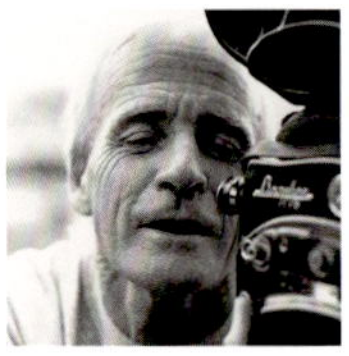 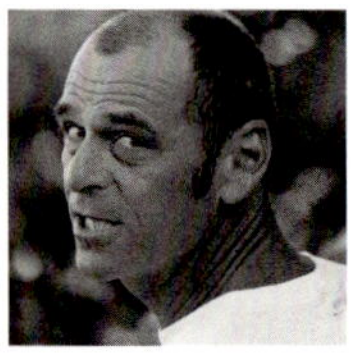

Dick Hoole & Jack McCoy

My first collective surf experience was at the Catamaran Hotel watching *Stormriders* in their theatre. That was the place I first heard group hooting, saw beers at a movie, smelled pot smoke...everything.

As I got older, I would never miss a Jack McCoy film—*Bunyip Dreaming, Green Iguana*, all of them. I finally got to spend time with him, and we got along great. He's had the privilege to document the best surfers in history. That's because his work towers above the others. Surfers want to film with a guy with that kind of a track record.

Dick Hoole was my avenue to old surf films. He's like a librarian. If I was looking for surfboard designs or wanting to check some obscure old surfer's style, Dick would source the film from his home in Australia and ship me a copy on tape. Not only that, but he'd tell me everything about that specific film I wanted: who made it, how much it cost, what it meant, if it made money, if it tanked. I've learned so much about the history of surf cinema thanks to Hoole.

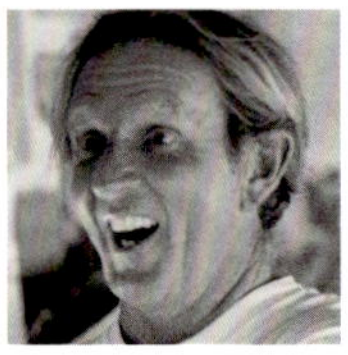

Skip Frye

The ultimate cult idol. He has a *following*. I had respect for him at 10 years old. From La Jolla south to the Cliffs; you know who Skip is and you can't help but have respect for him. It all comes from him being a purist. He never changes. Sweat pants, Converse shoes, and a fanny pack on the side, cruising around picking up litter at his beach. When was the last time you saw one of those kids at Pipeline picking up trash. Skip just cares.

He has no ego. As long as he can ride 11-foot boards and surf seven breaks from Luscomb's to Newbreak, that's all he wants from life.

I went to Costa Rica with him last year, and he has that same initial stoke that a kid has. To be 60 and still get amped up over a new board...that's where you want to be.

His surfing emulates a seagull. He takes the surfing of the early 1900s and blends it with the styles of today. When he rides, it just pours out of him. Even the way he walks on land, he's always avoiding the crowded route, reclusive and quiet.

His boards are 100% designed to ride the way he likes to ride. When you buy a Frye, you want to surf like Skip. He's still selling boards based on his surfing. Not many young surfers today order boards based on the surfing of the shaper. With Frye's, that's what you do.

You could put Skip, Kelly, and Curren in the water, and if someone who had never seen surfing watched the session, they'd be most impressed by how beautifully Skip surfs. Guaranteed.

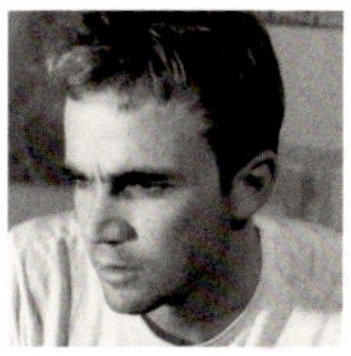

Thomas Campbell

I first got hipped to Thomas on a trip to *Thrasher Magazine* with Stecyk, up in San Francisco. I saw some posters he'd done on a layout table. I stole them and put them on my wall. So I was a fan of his artwork back when I was 16, and then I always saw his photos in the skate magazines.

Later, I met him at Cardiff. He was sitting under the lifeguard stand reading a book. What I like about T-Moss is that he finds beauty in subtle places others don't even look. In the longboard surfing realm, the finest elements are the most subtle, and Thomas got right into that. In surfing, it's the eye candy we see every day that we can become oblivious to if we're not perceptive. The offshores, the birds, the lonely moments are why we surf; Thomas gets all of that. Not the noise or the performance, but the beauty.

He has a huge ego, and thinks he knows everything. Artistically, he's not at all interested in any ideas but his own. He doesn't care about trends or movements. He does have an eye for style-conscious people: surfers and skaters both. He's not a groupie, though. He doesn't suck up to surfers. He could care less. Surfers are just tools for him to channel his view of surfing.

From his record label, to his gallery shows, to living for a few years in Morocco and Paris, he's a unique character. He treats his camera gear like garbage. Most photographers and filmmakers are so precious about their equipment. Thomas just rolls his stuff up in towels in the trunk of his car. If he gets a scratch on the lens, he says it helps the film.

He lived in this hellhole out in Brooklyn, pissing in bottles because he didn't have a bathroom. He's raw. He's no frou-frou artist.

His film, *The Seedling*, channeled the longboard vibe in a way that was really needed. Most of the longboard media was pretty lame to that point.

He could not care less about who's influential and who can help him and all of the high school social stuff. He's on his own trip.

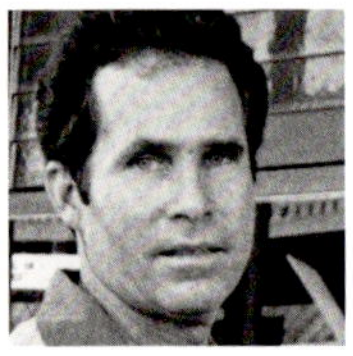

Duncan Campbell

It would really be nice if Duncan and his brother Malcolm got the credit they deserved. They devised the most far-reaching design in surfing...the three-finned surfboard. I don't know why, but the surf magazines seem to get their facts screwed up. The bottom line is that the Campbell brothers invented the three-fin. They're so humble. Talk to any of the pro surfers from the '70s that rode Bonzers- they know the truth. Brad Gerlach introduced me to them in the '80s...he was clued into it before me. Duncan always loaned me boards in Hawaii, and I love the way they combine the best of the single-fin feel with that of a typical, all-fins-the-same-size thruster. I'll say it again—the Campbell brothers invented the three-fin.

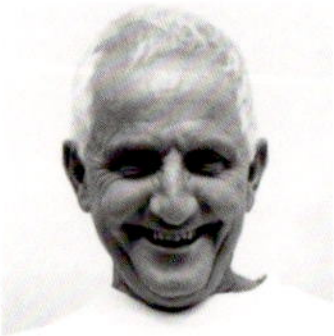

Bob McTavish

Applied science. He took horrible boards, hacked them down, messed with them, and made them worse. And didn't care. "Tomorrow's another day." He'd go home that night, rework them, and be back out there the next morning. He's so fired up on design, even today, that it's ridiculous. I mean, his energy level is just over the top. You tell him, "I've got this idea..." and he screams, "You've got an *idea?!* I've got a blank! Let's go!" Pure energy. He must have been so fun to work with in the '60s, with Nat and Greenough. Picture McTavish watching Nat spin out at Bells on one of those hideous V-Bottoms, and just taking a saw to it.

Nowadays it's, "Oh, my FCS fin broke...can I have one that's 3.5" bigger," or "My rocker's a quarter-inch off."

Surfing with Bob these days is a look at constant involvement: he goes through a quiver of three or four boards most days, trying each one out. And he'll rap to anyone about design. He doesn't care who you are, as long as you listen and are interested.

Life's full of takers. McTavish is a giver.

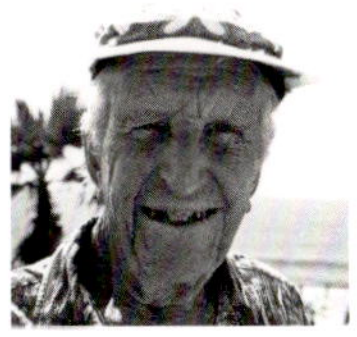

Woody Brown

WindanSea legend. First guy to jump off the cliffs in a homemade glider. Built boards. Woody Brown still surfs at 90. Think about that. He's seen it all, from the Duke to the passing of Dickie Cross, to today. Surfing kept him stoked and alive. Every day in Maui he walks out of his apartment and paddles out to the Wharf. He had Hawaii at its best. He'll happily talk to you and answer any questions. He's sharp as a tack, and has a great outlook.

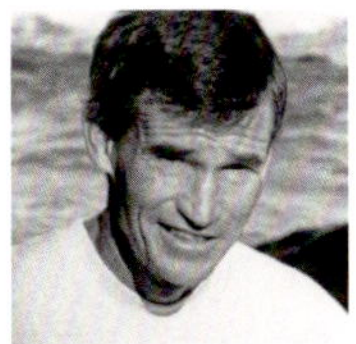

Jock Sutherland

People always used to tell me that I must really look up to Jocko; that we surf alike. I never really knew him. I just studied his articles in old surf magazines, and I've always had a thing for surfers that had a connection to the cosmic. He was informed by '60s culture, and he was the essence of that time. His surfing was color inspired. The things in his very experienced head—things he saw when he was experimenting—really showed in his surfing. The switching stance, the positioning, the risk-taking, and the different lines...he was feeling it. I know where he got some of that. I can relate.

He benefited from the positive elements of the '60s subculture, the attention to staying fit and being on a program. That's why he's still so good today.

In 1969 Jock was the surfer of the year—the craziest, most progressive time in our history. And he's still surfing well now. A lot of those guys aren't around any more...they're burned out or schizoid. He pulled it. What does that tell you?

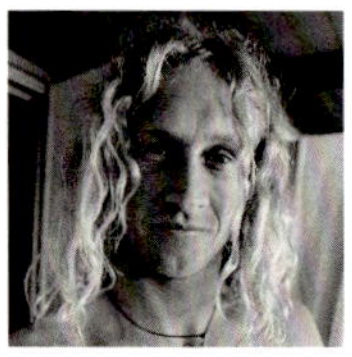

Rob Page

He used to sit and watch me eat when I was young, and he'd say, "Geez, grom, you're just putting shit in your body aren't you?" He might look like Jeff Spicoli, but he's incredibly fit and nutrition conscious. He was the first person to turn me onto whole foods.

He went through absolute hell following his Japan arrest, and the whole world wrote him off. There is a lesson there.

Most importantly, when I was 17, we were on a trip together and while watching a surf video I said, "God, look how old that guy is." Like the surfer should give it up or something. Pagey said, "Grom, one day you're going to be thirty and if you hear a kid say that you'll want to tell him to get fucked." Of course he was right. Age is nothing. Look at the guys in this book.

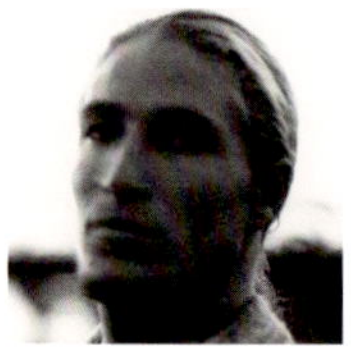

Beau Young

His dad was a father figure for me, so it's natural that Beau and I spent a lot of time together. Beau tried the pro shortboard deal for a while on the World Qualifying Series, but finally gravitated to longboards. Not easy. I mean, his dad's a five-time world champ, as well as...how do I say this... a very confident man. I'm sure things weren't always easy for Beau in that respect. Nat takes up a lot of oxygen. I felt weird because I was traveling all over the world with his dad, and that had to be strange for Beau back in Australia. Fortunately, by gravitating to longboards, Beau was able to share it with Nat, and it probably brought them closer together. But then again, when you enter Nat's arena, you're subject to criticism no matter who you are. Beau and I have that in common, so we connect on that level. We're friends.

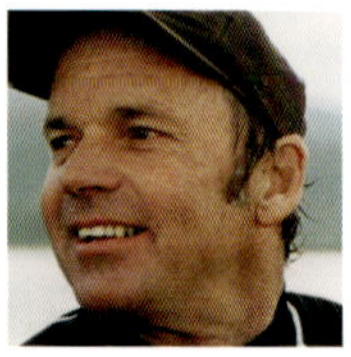

Billy Hamilton

Watching Hamilton in movies, he always struck me as a really polished version of Phil Edwards. Great arm positioning and poise, like a matador. Any surfer interested in straightening out an ugly style would do well to study Bill. That whole Hynson Red Fin crew—Skip Frye, Mike, BK, and Billy—just had a presence. And every single one of those guys transitioned to shorter boards with style.

There are a handful of people from the longboard era that I've just flat stolen from...Hamilton, I'm not ashamed to say, is at the top of that heap.

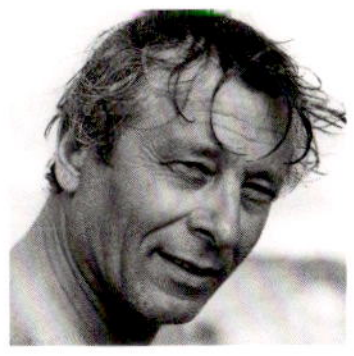

Herbie Fletcher

He stays active and he stays associated with youth. That's his secret, and that's why he's more on it than just about anyone else his age. And he's been there, he'll be the first to tell you. He predicts the future, and people don't want to hear it, then it comes true and they forget he called it. He doesn't care. He knows it.

Aerials? He said they'd be the future. Tow-in surfing? He was the first guy to tow someone in. Traction...not too many guys go without tail pads these days. And he's the only guy surfing Backdoor in his age group. "Where is everybody?" he says in the water. "The only people out here are kids!" He breaks the age gap.

You can't tell him anything, because he's been there and done that. And his opinions matter because he's done it personally.

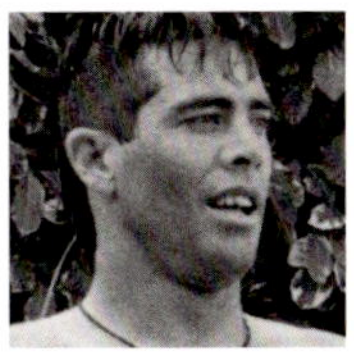

Christian Fletcher

He was punk, he was new, and he had flair. He was the first guy to focus on surfing in the air. He got written off severely. Now that kind of surfing is status quo.

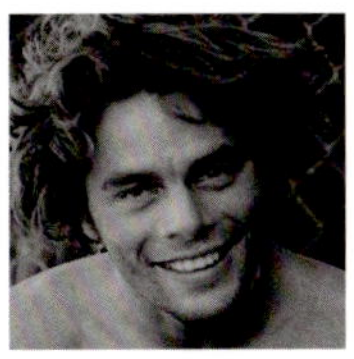

Nathan Fletcher

Overall freak, ability wise. In surfing today, everybody is so bland and flavorless. Nate has his own program, and he's been around a lot longer than people realize. He dropped out to race motocross and to be a professional snowboarder, and he's back harder than ever. He's a very artistic kid, like his mom Dibi.

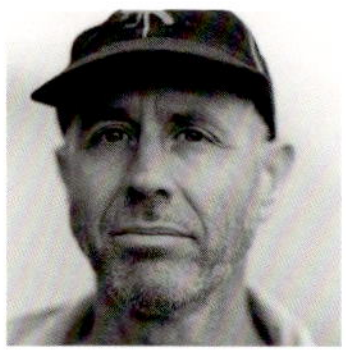

C.R.Stecyk

Now, Craig claims that I met him when I was 11, but I suspect that's just one of the BS roads he likes to drag people through. Which is one of the cool things about him. You never know what to believe, and everything is kind of like gamesmanship. Never bland. And like Hoole, he's an encyclopedia of surfing, and I don't mean trivia. He's constantly spitting information. He loves to trip you out.

He and Dora were supposedly close, but I saw them walk right by each other and never acknowledge one another at Malibu. I'm guessing that was some unspoken communication going on there. Part of the act.

When I was hanging around with him at contests, I had no idea that this big gangly guy with glasses was the man behind Dogtown and *Animal Chin.* My parents used to ask me, "Who's that weird guy hanging out with you." That weird guy turned out to be pretty influential for me. He always makes sure I have my facts straight.

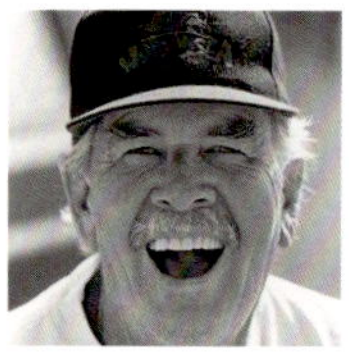

Mike Diffenderfer

For my generation, Diff was always a barely remembered name. It might have started when Surfer cropped him out of the frame for that Phil Edwards drop-knee sequence. But to hard-core big wave riders of the day, everyone wanted a Diff. The early North Shore guys knew his boards worked, but he was never a promoter. He seemed to lay low.

I was able to get pretty close to Diff through the Hynsons. He was the one who stressed the importance of heavy boards, which was so counter to the thinking of the early '90s.

At a holiday dinner, he took a half-full water glass and slid it across the table towards me. Right before it came to a stop, the sloshing water surged it forward a couple of inches. That was his momentum exercise. "Don't write off heavy equipment" he said; "there are benefits there."

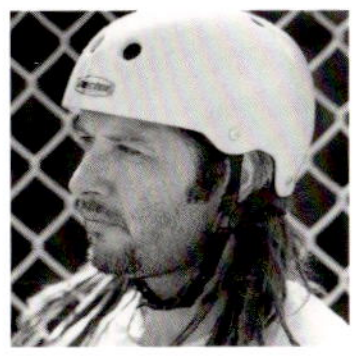

Tony Alva

Thirty years of pool skating. You grow up in The Valley—the suburbs—you skateboard. I skated before I surfed. All of us did. You know who Tony Alva is. Seeing him in Super Session and those films...the fades up against the fence on those banks, the style...his style is just priceless. The lines he draws on waves and in pools are just beautiful. The way he links lines and turns, the way he covers distance, sets him apart. Now he's 45, and he's still doing it. Most people who skateboard for a long time end up crippled. Falling hurts. Concrete isn't water.

His style and finesse have kept him fairly injury free, and when you watch him carve, you get chicken skin.

I went to the new Coronado skate park with him, and when he walked up to the bowl he said, "Move over boys, Big Daddy's here." Everybody pulled back and was just hooting as he dropped in. It was like surfers seeing Tom Curren at their home break.

Most skaters don't give surfers the time of day, but Tony's a surfer first, and he's always been willing to rap. A sad thing is that the surfer/skater is a dying breed. There are a lot of posers who dress like they skate, and they take skate moves to the water, but they can't even drop in on a ramp. The two disciplines do a lot for each other.

Today his life is about finding pools, riding waves, and selling skateboards. Same as ever.

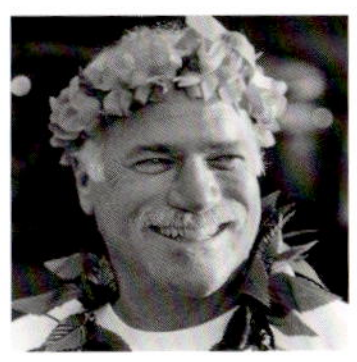

Mark Martinson

I admire Martinson. It's so funny to see him in the old MacGillivray-Freeman films, all innocent-looking and always smiling. It's funny, because at that time he was running for his life from The Man. He was dodging a phony war. Surfing, and traveling around the world, and having it documented by a couple of great filmmakers. What's not to like about that?

I really like the way he rode longboards, especially in *The Performers*. And later, there were only a few guys who rode those transition era boards well. Martinson's body type and power was a perfect fit. And that scene in *The Sunshine Sea* when he comes through that tube section shaking his hands and claiming and hooting. That sums up the joy of surfing.

Best of all, that's how he is on land today—all happy and smiling and never bitter. He's just really carefree.

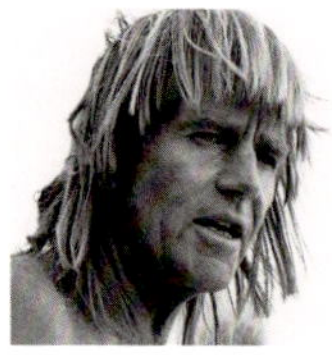

George Greenough

The first time I met George was with Beau Young and J Brother. We were making surfboards; our first attempt at the "do everything" approach, shaping, glassing, sanding and all. We ended up trying to track Greenough down to get help finning them. Nat told us to go to Broken Head and look for the pyramid. Well, it wasn't easy because Broken Head's a national park and there's a whopping seven houses out there. We drove up random driveways into the hills, where we'd be confronted by people asking us what the hell we were doing. They'd heard of Greenough, but that was it. Finally, Beau saw the pyramid and we made our way up his driveway right to his doorstep.

We started hollering, "George...George!"

Out he comes, wearing a pair of the dirtiest shorts you've ever seen in your life, no shoes, no shirt, hair cut with a knife...exactly like we expected. "Boy, you're Aryan looking," he said to Beau. He hadn't seen Beau since he was Nat's baby boy. After that he just took us in. We started drilling him with questions and told him we wanted to make some fins. He laughed and said we weren't going to be doing it there. "It makes a fucking mess," he said. But he did haul out some buckets of fins, and described them to us. "This one flexes like this, but not like this," and "This one bends here, but it won't do this." There were half-buried sailboards in the ground, scraps of metal, half-finished projects...it was crazy.

In the back was a 15-foot deep ditch lined with truck tires. He told us it was a rainwater irrigation system for his organic garden. He hasn't had to use a tap since he built it. That's advanced.

He let us check out his water housing from *Big Wednesday*. It weighed about 60 lbs.

He was blown away with how stagnant board design is today, especially among the pros. "These guys are makin' a couple hundred grand a year, and you can't invest 25 grand in board and fin design?" It just didn't make sense to him. Then he broke out underwater photos of tri-fins, and pointed out how the fins weren't working. We didn't understand him, because he was talking a mile a minute.

His current project, *Dolphin Glide*, is crazy. He showed us some stills, explaining that dolphins wait for special moments to ride waves; the most beautiful sunsets, the clearest water of the year, the crispest swell of the season. His contributions to surfing were just a phase for him. He still sailboards and mat rides, but *Dolphin Glide* seems pretty important to him.

The thing about George that is so great is that it took a kneeboarder to figure things out.

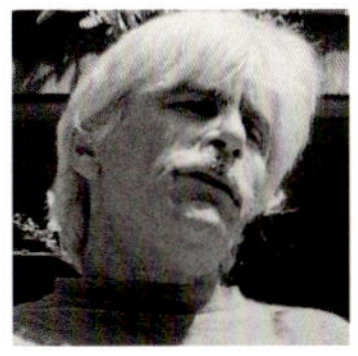

Tony Caramanico

TC is an example of how a hot, experienced local guy who's been around the block can teach you as much about surfing as a star; the most random person can be really influential. I met Tony when I was 16 at Donald's factory. He always told me, "You've got to come to New York." I'd been all over the world, and had no interest in New York. When he finally talked me into coming out, it was my first dose of jazz music in my life. A guy who rented from him in Montauk was the son of Ed Loch, a jazz drummer who had played with Coleman Hawkins and Ella Fitzgerald. Anyway, he had a crew of jazz musicians that sessioned in Montauk, and I got a real introduction to the sound...from the inside.

To that point, I'd listened to really bad music...I'm talking death metal.

The first moment I heard their band, it all made sense: the tones, the pauses, it all sounded like longboard surfing to me. The stalls, the footsteps, the note runs—it's all identifiable.

Next, through Tony I was introduced to the Montauk and New York art world. For example, I had no idea that the house we were tripping around was Peter Beard's until I studied up a bit. Getting to interact with fascinating people I would have never otherwise met—that was all due to Tony.

From a surfing point of view, I really like how the Montauk people enjoy the beach. They only have three months to be out in the sun surfing, and they make the most of it, it's an amazing vibe, and nobody takes it for granted like Californians do. It's my summer retreat, and I haven't missed a season there since.

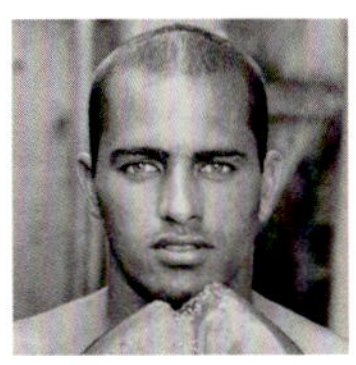

Kelly Slater

The champ. It's impossible to knock him. He's had to overcome so much hype and pressure. The accomplishments he's racked up before the age of 30? Coming from Florida? You've got to be kidding me. He changed the way tubes are ridden backside. He changed the way Backdoor is surfed. I learned how competitive he is over a game of pool in Lacanau.

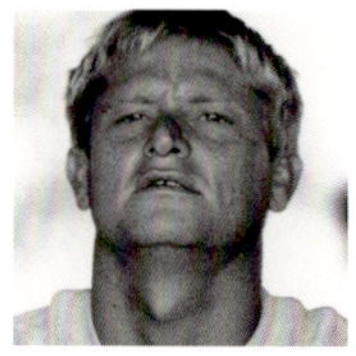

Strider Wasilewski

Strider was my introduction to how to survive on the North Shore. He was just about the only guy from California surfing Pipeline when I started, and I learned a lot from his example.

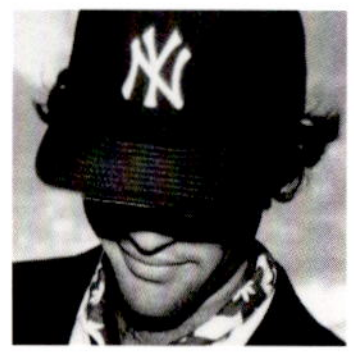

J Brother

Joe's an accidental filmmaker. He doesn't plan anything. It takes him five years to make a 20-minute film. He's never had to face the market realities of "We have to make the deadline for the Christmas video sales." His whole trip is built on trying to make something timeless. His short, *adrift*, raised the surf film bar in a quiet way, and reflected the longboard trip in a complimentary way.

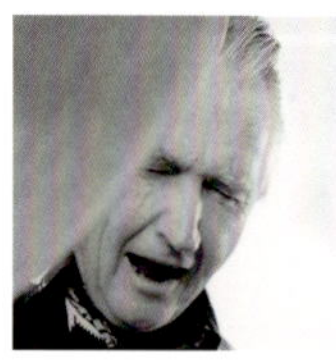

Tom Morey

The spaceman. Some day a spaceship is going to come get him. Every time you see him, you know you're in for some knowledge. He never talks about himself—only projects. We connected because of jazz. His tales of Dizzy Gillespie and others he played with are great, and he has that jazz cool about him; he's just always high on music.

He has such an open mind where design is concerned. He's tuned my boards before; put duct tape edges on them for me.

You'll go to his house, and he's working with magnetic levitation and probes, then he'll break free and sit down at the drum kit. He's making all of these great faces, and it's like he's surfing.

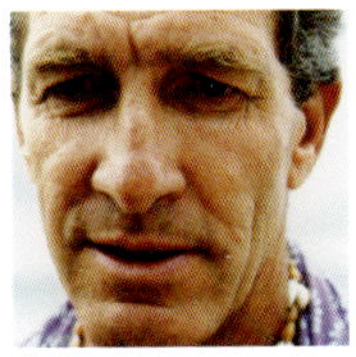

Nat Young

I met Nat when I was 12, and at 14 I had to compete against him. That was at the Oceanside Pier. Halfway through the heat I thought I was doing him up, but he was just waiting. He taught me patience. He was in his fifties and he was still the best of the best on a longboard. Beyond that, Nat was the man. Yeah, there were guys that were beautiful—Hamilton, David, Jock—but Nat had the build to really power the equipment back then. Greenough told me that. Only Nat had the physique to throw those old pieces of shit around. Mid '60s boards were built to nose ride, but Nat was the first guy to really throw them around; not little half-turns either. Real turns. You can see that in movies like *Hot Generation* and *A Winter's Tale*.

He was the first guy to actively pump a surfboard. You just know that Bertlemann saw that and said, "Yeah, knees above the chest, the springing bottom turns, flying, that's for me." There were guys that rode the nose way better than Nat, and that's a fact. But Nat broke trim and went places.

Nat's super passionate about surfing, decade after decade. Fins, for example. I'd show him a pile of my fins and he'd roar, "Boris, these fins are shit-house, come here..." and he'd be cutting pieces of paper to shim them in the box so they wouldn't move a millimeter, or he'd dive in and start wet-sanding every bump away. He emphasized keeping a clean stick; cleaning any speck of wax off the bottom and rails, polishing it up before a heat...he was so into it. He had an entire regimen of waxing, just the right way. Nat works under a set of strict principles.

He taught me how to travel, how to really experience foreign culture. None of this leave the contest and hide in your hotel crap. Go out at night, dive into the cuisine, the wine, the people. One evening I was chewing gum at a dinner reception, and I shotgunned a glass of champagne. BAM! He clocked me in the back of the head with an open hand. "You never slam champagne, like a woman—you enjoy every moment of it. Don't you have any manners?" Nat's always been into the finer things.

Traveling in Europe with him was so funny, because he'd been going there since the '60s. He was like a diplomat. He taught me how to be a gentleman in certain situations and how to be a showman, from a personality point of view. He walks into a room, and all eyes are on him. Same thing when he hits the water. We'd go out at Bells and he'd just take over. A lot of people resent him for that, but he has juice. Total command and total charisma. And, it's fair to say, total arrogance. He's almost killed me during big wave sessions. He did that to show me I could do it. He likes to test people's courage in the water.

One thing though—he's the most competitive person I've ever met. Every surf you have with him, it's like a horn went off. His confidence is outrageous. He's larger than life.

The whole thing with his getting his ass kicked at Angourie...he's a bigger person because of it. But he's still passionate; he's still traveling the world, skiing and surfing. He still does his diary every day, his yoga. His mantra is "keep moving." Don't ever sit down. Don't ever stop. He comes unglued if he isn't moving. He'll never stop.

When I'm heading towards 60, I hope I can maintain that passion.

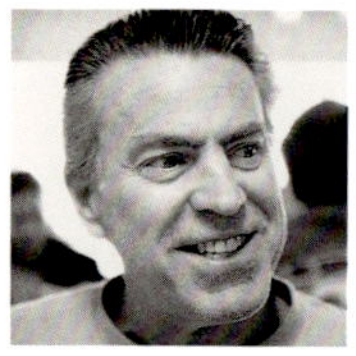

Lance Carson

Try as I might, Lance was never really accessible to me. There were times I'd be surfing Malibu and he'd walk all the way up from down the beach, through the pier, paddle out, ride two waves, and then disappear.

The footage of Lance in *The Endless Summer*...he was the start of the longboard bloodline from where I stand. He made me want to ride a longboard...him stalling and the wave's hitting him in the back, and he's just holding it there in the pocket...whew.

For us longboard kids in the '80s, he was the beginning. Footage of him showed us how to surf.

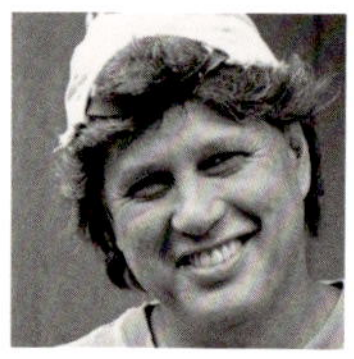

Barry Kanaiaupuni

One of the most unique surfing styles I've ever seen. That close, almost animalistic stance, and his total poise really set him apart. How about doing stalls at ten foot Sunset Beach? His positioning was out of control. He owned Sunset, and the way he rode Honolua and Velzyland was just as distinctive. He always looks so heavy in the movies, but he's the sweetest, most family-oriented guy you could meet. He transitioned from the longboard era to the shortboard era beautifully—from Mike Hynson Red Fins to those blades he made and ripped on. That stuff of him in Cosmic Children where he's on the 16-inch wide board...what can you say? No one can dupe that.

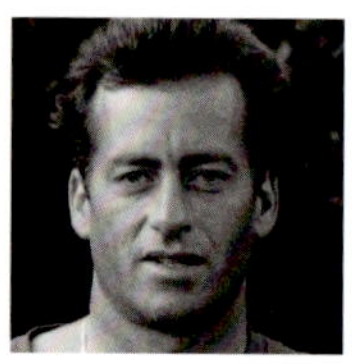

Tom Curren

The last non-commercial surfer? Who knows? Now, as soon as a hot guy comes out of the woodwork you know his whole life story six months later. It's like the Backstreet Boys go Surfing. There's a packaged vibe to so much today. Curren had some mystery, and he knows when to hold back. You can't get close to him; kind of like his dad. They both disappeared on their own terms. Via "The Search," he sort of invented the concept of the "photo surfer." A lot of people owe him credit for that; used to be you had to compete to get paid to surf. Stylistically, he just had the genetics. He was so soulful and smooth that he could do head-snaps and still be the most fluid guy around. Single-fin roots, you know?

Curren's going to just get more and more low profile, and I suspect that all we'll hear of him will be random "Curren Sightings."

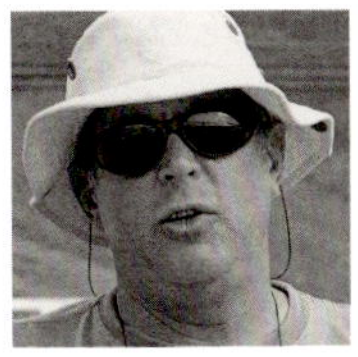

Art Brewer

Art gave me a chance when nobody else would. I'd mustered up the courage when I was 14, and asked Aaron Chang and Tom Servais if they'd shoot with me, and they just shot me down. Aaron said, "I don't shoot longboarders and I never will. I can't sell the photos." Servais told me the same thing. I was naïve back then. I didn't know how mercantile it all was.

About that same time, a cheap little free surf mag asked Art what he saw in surfing that was fresh or something like that, and he named me. I was blown away. I didn't know who he was, but it was crazy to read that. I called him up, and he said he'd shoot with me. Imagine that—a surf photographer actually pointing his lens at something that might not sell.

That really boosted my confidence, and when other photographers would shut me down, I'd say, "Well, I'm good enough for the Avedon of surfing to shoot, so screw you."

That always made me feel better.

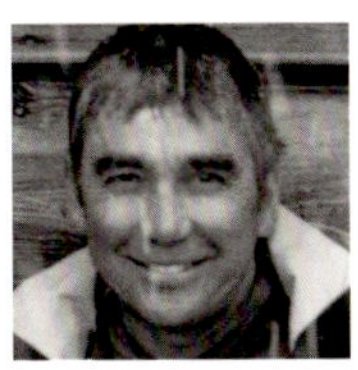

Gerry Lopez

My brother and I grew up with posters of Gerry on my walls, and got to hear the audience explode at surf movies when he dropped in at Pipe.

My first real glimpse of Lopez was in France, strangely enough. He was riding a longboard at Côte Basque, and that style of his is unmistakable. Later that afternoon we bumped into each other, and I was star struck. I tried to keep my cool, but it wasn't easy. My friend Takuji and I finally worked up the courage to ask him if he'd rent us a room at his house at Pipeline. We ended up getting the house that winter, and that became my annual North Shore home.

The second season there, I got my first real lesson from Gerry. I'd been riding single fins at Pipe. I was upstairs, and I saw Lopez checking out my boards down below. When Gerry is touching your boards, you're kind of..."whoa."

He came inside and said, "You *like* eatin' shit? Those rails, you must be sliding all over. I don't know how you can ride those. Come upstairs for a minute."

Well, "upstairs" meant his private room that was always locked up when renters were there. He walked me in, and started pulling down boards. He gave me a lesson in vee and rails, with his glasses on the whole time. He never takes his shades off, even indoors. "You want to ride Pipeline, this is what you want," he said. He showed me a simple formula: flat deck, slight nose rocker, vee from halfway back, super down rails, wing pin. It was the most beautiful surfboard I'd ever seen. I went home and had another shaper make me one to those specs, and it changed the way I rode Pipe.

Another time, I was watching the surf there, and I asked him if he was going out. He said no. "I had this place for 15 years to myself...it's hard to get motivated unless it's really good." Imagine having the best wave in the world for 15 years, living there on the beach with the best surfboards ever built. Those boards they had from 1972-'75 were so advanced. They weren't held back at all; perfect for the job at hand. No logos...it was as pure as it comes.

Lopez's most valuable lesson is in how he treats people. He's a complete gentleman, even with random characters that meet him for the first time, star struck.

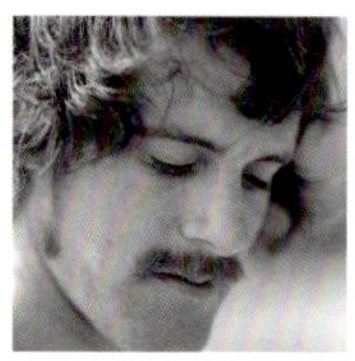

Donovan Frankenreiter

Donnie's one of the coolest guys in surfing. He adds welcome flavor to the pro surfing world. You can call him a hippie and say it's an act, but he hasn't changed. Everything about him is original. He collects surfboards to ride (he has over 200), before any of the other kids, he was working over the thrift stores across the nation, and he rides with beautiful style untainted by thruster-itis.

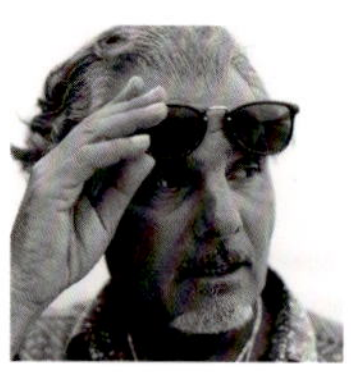

Miki Dora

Biarritz airport, 1992. I'm 15, sitting on my skateboard at the oversize baggage claim. And Dora walks up to me. "Oh," he says. "Aren't you the famous kneeboarder? Some kind of wave beggar?" What do you say to that? I was jet-lagged, just wanting to find a cheeseburger.

Later in the week, late at night, we shared a taxi from a function in Biarritz back to our hotels, and for the first few minutes of the ride he didn't say a word. He looked over at me, frowning, and asked, "Have you ever walked the path of the human being?" He didn't blink, and was just glaring at me, waiting for an answer. What could I say that would satisfy him? It made no sense to me. I just looked at him. After a very long minute, he cracked a smile and answered for me. "Pretty cockroachy, isn't it." He started laughing.

After that, we hung out a bit every day in France. The bullshit he'd feed you, the expressions, and the animated faces he'd make...I'll never forget a single thing he said to me.

His whole life was an orchestrated act, and he never broke character.

I saw him surf by himself one day on my board, and this photographer on the beach saw who it was. He was blowing up his mat as fast as he could, almost hyperventilating. Dora hadn't lost a beat...he was surfing perfectly, the hand motions, everything. The photog kicked his way out to Miki, and when he finally made it out there, Miki went into his "kook" act. Whenever the photographer would get him in his sights, Dora would slip off his feet and fall off. He did it five waves in a row just to dick with the guy.

He walked up the beach and handed me my board. "Biggest piece of shit I've ever had the displeasure of surfing." Last time I talked to him that trip.

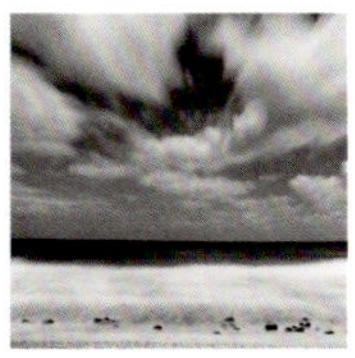

Hawaii

The ultimate surfing homeland, the ultimate proving ground, and the ultimate respect teacher...the ultimate experience. The benefits of pulling a winter there are incalculable. You get what for many is a first look at reverse racism. When you show up on the North Shore, you're an outsider in many ways. You can critique localism as much as you care, but in Hawaii order is maintained. Things aren't finished in a courtroom or through a lawsuit. There are clear rules and standards. The only way to get respect is to pay dues. You can't come and take. There's a culture you need to clip into. It's the hardest place in the world to feel comfortable if you're not respected. The surfers there are happy to enforce the rules.

By spending your winters and taking your beatings, you are eventually accepted. It's a serious accomplishment. It's like getting a black belt. It's based on how you carry yourself and how you treat people. It's about being humble, because if you're not you're going to get put in your place—fast.

Aggro Ambivalente

It's not one of those. Glossy gossamer ghostwritten celebrity tell-alls emanating from the gilded sands of the Grande Plage. No, this is something entirely different. What I'm supposed to do is provide a brief intro to these goings on.

Santa Monica. Shit, the monkey on my back at present is a meandering 15,000 words or so of research observations. Kept together in a pile, maybe more aptly it's just a pile. A question. How to produce anything beyond prosaic nomenclature á la "the surf-sliders-turned-artists Halsband and Joel wandered the trackless wasteland seeking aquatic seraglios of evanescent pleasure, scribbling notes and racking focus as they went?"

Break it down. The champion and the chronicler both exhibit a Hades-honed self-expression. What, when, where, why? Alas there the accumulation of data sits. Underlying assumptions generate more speculation and additional findings. Which in turn are biting my backside. Prodding me to further spin out other missives. Speculation in search of direction.

Just out of New York. As my first effort at summing up this miscellany I offered the quote: "Everyone makes mistakes, and God made Joel Tudor." They should have taken it and run. Knowledge and exactitude are but mere illusions, but still I fret. Perhaps too much? How to best serve this project? Chase validation via association by interviewing the champion du monde and Michael in Trump's boathouse on the lower Hudson? Go with the uptown, A-train rollin' country mouse goes to Gotham City hook? Wallow in the elegant ennui of two ballers on a self-imposed budget? Sink into the operatic abyss of the Rolling Stones Earl McGrath cheese stackin' in Manhattan link? Or succumb to the patriotic our gang hangs here homily as Tudor meets Kelly Slater meets Halsband meets Tony Hawk meets Uncle Art Brewer on the flight deck of USS Intrepid? Long may it wave.

Confucius advised that to understand the innermost poignancy of humor the acolyte needed to "follow the monkey and to forget how foolish one looks." Once I was lost, but now I'm a flounder migrating to Sheepshead Bay in the hopes of encountering my simian muse. Drew Kampion the poet laureate of surfing is in town. If there was ever a man. For this job. At these prices he'd produce triplet couplets on the spot and eat the afterbirth. Me, I'm at my most bitchy up in the club. Down but not out. I fumble through a walkabout in a pallid attempt to be near the impact zone in the unlikely possibility that the phantom muse will perform a HALO (high altitude low opening) insertion into these proceedings.

Ensenada Baja Califas. Weeks have transpired, so for the good of the cause I find myself holed up on the Mexican coast. Here, Doctor Carlos Gallegos Rivas graciously allows me to use a room and his power outlet. The good *dentista* believes in the power of narrative and understands the finer points of humanistic enterprise. For that reason, south of the border, I drift back in the beginning:

AUDIO; IMMEDIATE BACKGROUND, RAPIDLY INCREASING IN AMPLITUDE:

"Fuck O Spastic, weren't you a grom once?"

AUDIO; CLOSE RANGE, THE SOUND OF BREAKING DISHES.

Robert "Gnat" Young was taking me out from the blindside. The Animal, as he is generally referred to, likes to give better than he gets. In this instance he was howling like a wombach in heat while he flew over a railing and two rows of tables to achieve his objective, which was knocking me to the ground hard. But the best of it is that his antipodean antics shocked the shit out of those Blue Point oyster sucking yacht clubbers. Nothing's sacred and Nat will later author a book on "Surf Rage" in which he will ponder the psychological implications of thuggery such as the two of us are engaged in. But for now he and I were rolling and tumbling and pushing and shoving and chortling hysterically over the childish exuberance of these things. Celebrations of first fleet convict chic hold nothing sacred, and in a polished mausoleum of conventional propriety such as the San Diegito Yacht Club, a mate has got to both protect and to serve. Young and I were marked men in this occupied territory because we had no burgees, blazers nor manners. The wreck of the good ship Dennis Conner was sitting off to the side looking decidedly dejected. As well he should have, because the Commodore knows that Nat can walk on water. And that ability invariably wins friends and influences things. Unfortunately the hellhound phantoms of Ben Lexcen's winged Aussie keel and the golden hulled Kiwis had already chased America's greatest skipper onto a spiky precipice on the shoals of respectability. This dry reefing had reduced Captain Conner to the ignominy of campaigning a 60' catamaran in the most transparent of Cup subterfuges. Dennis knew everything was over when the last harbinger of doom sauntered through the front door. His name was Joel Tudor and he was fifteen years old.

The boy wonder of surfing was wearing no shoes or shirt and he was looking for fun rather than service. Consequently we were shown the portal and asked to leave. The Animal rained down science on his hirsute protégé: "In a place like this, always order off the menu, send wine to the chef, eat fast, tip big and leave quick."

We walked down the anchorage slips with Young entreating me to "give Boris the Clitoris a break." Nat was speaking as a tribal elder, I gathered it was some sort of expected duty for me to render support of Tudor's preciosity. Young's use of Joel's Australian sobriquet was more a commentary on the youth's pink wetsuit affectation than it was affectionate. My attention was directed elsewhere when a comely wench slunk over to the railing of her racing sloop, squatted down and began talking to Nat. It was apparent that she was a fan. Joel and I stood off to Young's side and marveled at his game. We three were standing on the dock while she was straddling the rail wearing a silk dress sans underwear. Tudor reckoned that she must have been French. (Even at that relatively early age Joel was already a European Grand Champion.) I recognized how ambassadorial and diplomatic Robert Young the four-time world surfing champion had become. Nat demands to be taken seriously and has run for elected office in the past. Plus he stands at six feet three inches tall. When you factor in the elevation of the boat and calculate the angle of the Animal's view, and add in the skillful stealth of our hostess showcasing of her assets, one can begin to appreciate the way that fundamental public relations work. Joel and I slipped away to our Bible studies leaving Young to line up votes and to be lionized by the adoring legions.

The totality of these aforementioned proceedings demonstrates that our authors did this treatise because they could. But more importantly they did it because they had to. Joel Tudor embodies the lightness of being which descends upon the perpetual voyager as they watch the continent drift irretrievably away over the horizon. Michael Halsband is unhinged and unfettered in the way a witness is whom has seen too much. This melange is the irrefutable result of the teaming of such survivors. Tudor worked his way through the zarathustrian drama of competing for a decade on the world pro tour circuit. Halsband endured the pandering of a thousand fahionistas as well as the stuttering metapausal sang froid of the Cirque de Stones. Because of these things we end up with virtuosos obsessively inclined to document only that which interests them and whom have the social chops and financial clout to get such a meandering miasma funded.

Michael's first take with Joel was the defiantly anti production "Surf Movie: Reels 1-14." 48 minutes. Uncut. Unscripted. Unenhanced. A vérité approach that was revolutionarily deconstructivist amidst the overproduced advertorial product that passes as surfing documentation these days. The prolonged collaborative effort of Surf Movie continued into this title, which is by its very existence an implied commentary upon the absurd state of affairs that surround contemporary surfing.

Surf Sells. Detroit Autorama. Toyota flogs Roxy Echos. Kia designs Cinco Swim show cars with neoprene seat covers and built in sand washing nozzles for apres surf. Porsche displays built in surf racks. Ford SUV's are pitched by the glaringly incongruous "longboarder" LA Laker Rick Fox. Manhattan Island. Hang Ten polycotton separates are racked at Bloomingdales. The New York Times covers "surf fashion" with the paper of record dubbing it "A Shore Thing." Tommy Hilfiger hangs boards as prêt-à porter props in his showroom. The Wall Street Journal notes that a Times Square surf shop sells neither surfboards nor wetsuits. Sepang, Kuala Lumpur. Fair dinkum surf murals placed by Fosters Lager exhort Formula 1 Grand Prix guests to "blow the foam off the top of a few." Paris. Gucci bitches strut with surfsticks down the haute couture runway. Seattle. Amazon.com offers the "totally groovy, totally retro 1971 Malibu Barbie" on the World Wide Web, Malibu. Tourists drive rented Chevrolet Malibus that are assembled in Kansas. The Coffee Bean and Tea Leaf shop has a 6'10" Phil Becker surfboard autographed by such presumed provincials as Weird Al Yankovic, Tony Orlando, Cindy Crawford, James Cameron, Brian Grazer, Reggie Miller, Fonzie Winkler, Dick Van Dyke and Vanessa Redgrave. Huntington Beach. ESPN-Disney-ABC unveil the X Games team surfing format. This Orwellianly Oedipal grafting of TV controlled team sport onto the appendage of the obligatory extraneous boob shot injects new kink into the big brother- prodigious mammary lexicon. Hollywet. Coors erects three dimensional neon lit thruster snap backs alongside Interstate 405. James Bond appropriates tow in. Charlie, Bosley and the Angels go surfing on screen. You see, Cameron Diaz "really" rips, just check out her atrociously retouched billboard faux tube-ride for proof. People mag shows us "actual" surfers like Adam Sandler while the Los Angeles Times serves up a wave riding Rodney King at Humaliwu. Such happenings are demonstration that yesterday's dead beats and beat down's are today's new Gidgets. Thus ingenue progenitors of the pop culture paradigm are sickeningly surf chic.

Can it sink much lower than this? Missionary moralists at Esquire have even published a dissertation on whether former United States amateur surfing champion Malia Jones was sexy. Should her fellow Hawaiians thank the gods now that Polynesian dress and mores are finally acceptable practices on the eastern seaboard? Can it really be any surprise that a super model that's also world class competitor is considered sexually alluring? On Manhattan Island apparently this is news and represented such a cerebral stretch for Esquire that they needed to further rhetorically cogitate "Is it any wonder she's a Woman We Love" If it wasn't for surfing's new found acceptability we might never have read of this anomaly.

Riding of the *papa he`e nalu* is a sport/activity with several thousand years of heritage that now finds itself a victim of the intellectual elite's moribund attentions. Will this ancient practice be able to survive being *au courant?* May rampant institutionalized appropriations of surfing imagery eventually neuter the sport of Hawaiian kings? Philately's new trend tests the extinction via overexposure assumption in such far-flung governmental stamp issues as: The Caicos Islands edition of Walt Disney's Goofy tube sliding. The Christmas Islands fur trimmed surf Santa Claus. French Polynesia's Aerielenne postal wave turn. And Washington D.C.'s pressure sensitive adhesive backed Duke Kahanamouku USPO 37 cent commemorative. Where do we go from here?

The protagonists in the tale at hand on occasion index with Diesel and shill for GAP. What are the author's motives in this publication? A repudiation of their own past spiritual infidelities? The testing of a hydro hypothesis? The cataloging of natural science aberrations in remote habitats? This book isn't an outsider narrative like Edison's 1898 movie of Hawaiian surfing, which constituted the world's first sustained view of the act. It's not a pseudo ersatz-surf homoerotic catazine being propagated by some grifting Ohio schmata men. Nor is it a miscreantic recording of the heroic spectacle of sport. This is the musings of a couple of guys as they ramble through their own lives. Surf semiotics aside, this documentation raises questions of a metaphysical nature. Knowledge is an instrument of action, not some theoretical representation. When the American philosopher John Dewey proposed in the late 1800s that art was about experience and not mere ideas he was only a couple of thousand years behind the islanders. Polynesian pragmatism had long understood that surfing was an integral aspect of a being's totality and was thus absolutely inseparable from cultural/religious/educational/utilitarian concerns.

Specifics on this project? My nephew Joel showed up one morning, out of the blue, on a mission with Michael. I didn't need to know, so I didn't. Whatever they asked me to do I did, and off they went. Another episode gone. Until a year after. When the proofs dropped one day via Fed Ex. I looked and realized by the by that I pretty much was connected to everybody else in the book. This was an unsettling discovery to someone some one such as myself that had been inculcated with the canons of the ruggedly individualistic *cogito ergo I ride sect*. What does this reveal about the denizens of our so-called solo sport? We are more alike than we ever care to admit. All being linked by the unavoidability of our choice to partake in this activity, our sense of how to live and the inevitable recurrence of this shared urge to glide. The best friends any of us will ever have are the other dwellers on our small island.

Joel and Michael are allowed access to moments of privacy. They are trusted to get it correct because everyone knows that if they don't self-regulation will take place later. The potential horrific consequence of being jumped out of the clan keeps things honest. For that reason value this view, as you aren't going to see it elsewhere.

El Pueblo de Nuestra Señora la Reina de Los Angeles de Porciuncula. I spent the other day downtown in the Superior Court of the State of California. Hung up watching some one else's problem. Listening to a judge and a couple of lawyers try to figure who did what to whom and who sold off what and where is the money. Therein dwells the true price of justice, as the legal system is entirely dependent upon compensation. If there's no pay the barristers won't play. Hence the judge solemnly imparted "This court finds it impossible to believe that any surfboard is worth the time and expense we are lavishing on this." Now it should be glaringly obvious that the judge doesn't surf and the lawyers are kooks too as each of them clock and cluck for dollars. But a magic board is gone and it was a gift at that. The totality of all that rightness of form, discriminating proportion and faultless balance will never come this way again. That is a real crime, and anybody that appears in this book will tell you that. Functional art is hard wrought, as utilitarianism is a prerequisite for existence in relational aesthetics.

This is Joel Tudor and Michael Halsband's treatise. About those who walk on water. Which is as good as it ever gets.

– C.R.Stecyk

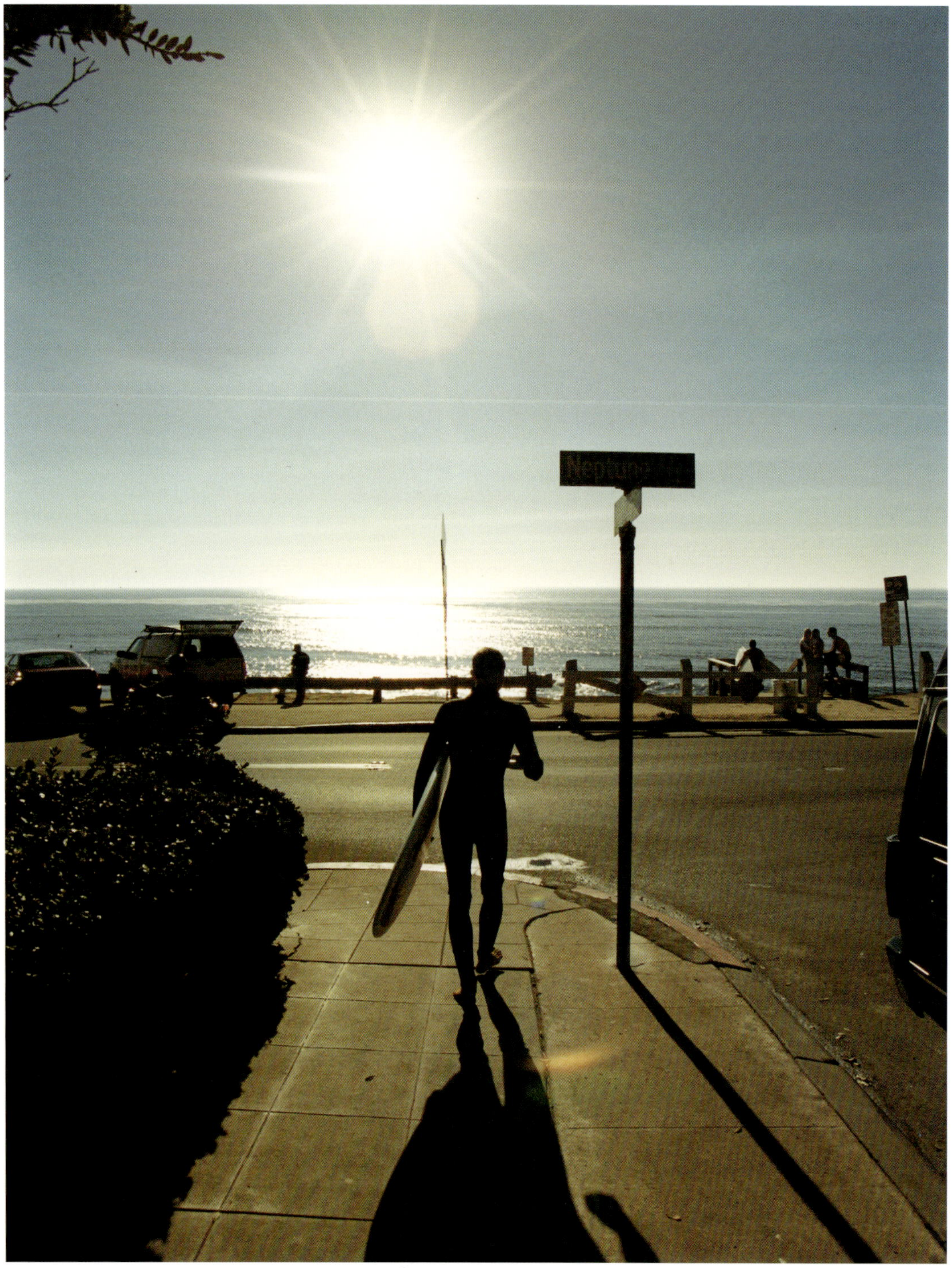

Joel Tudor Acknowledgements

My dad, my mom, my brother, Maya Brown, Uncle Tony,
Donald Takayama, Nat Young David Nuuhiwa, Gerry, Brett, Stretch,
Devon Howard, Mitch's Surf Shop, Joe Roper, WindanSea Surf Club,
Greg Noll, Michael Peterson, Tom Wegener, the Horseshoe locals
(especially Rolf and Haffie), Ben Murphy, Skip and Donna Frye,
Jackie and Josh Baxter, Darren Ledingham, Alex Cox, Lance Ho' okano,
Dino Miranda, Rory Russell, Tom Stone, Dru Harrison, Chris Slickenmeyer,
Izzy and Jonathon Paskowitz, Bill Dice, Stu Kenson, Chuck Hasley,
Tony Channin, Tom Eberly, the Bahne family, Richard and Peter Kenvin,
Peter King, Tom Andrews, Nathan Jernigan, Matt Rimel, Brian Surrat,
Jack Luber, Stu McRobbie, Terry Ahue, Bobby Challenger,
Mark Anderson, Reese Patterson, Robert Weaver, Takuji Masuda,
Michiaki Ishida, Lisa Cooper, Chris Ahrens, Sam Ryan,
Malcolm Campbell, Guy Takayama, Oceanside Longboard Club,
Jim Irwin, Jeff Kramer, Joey Hawkins, Josh Farberow,
Jed Morouse, Lance Woleslagle, Dylan Jones, Scott Stecker,
Hynson Jr., Melinda Merryweather, Rex Huffman, Eric Huffman,
Jason Murphy, Jeff Junkins, Bear Mirondan, Miles Davis, Jeff Lock,
Big Ed Locke, Ed Lock Sr., Reese Simpson, Skip Engblom,
High Speed Productions, Think Skateboards, Independent Trucks,
N.H.S. Rich Novak, Anti Hero, John Coltrane, Steve Lis, Peter Hewitt,
Bruce Lee, Rodrigo Meidiros, Kai Garcia, Phillip Castognola,
Jesus Christ, and the rest who know who they are.

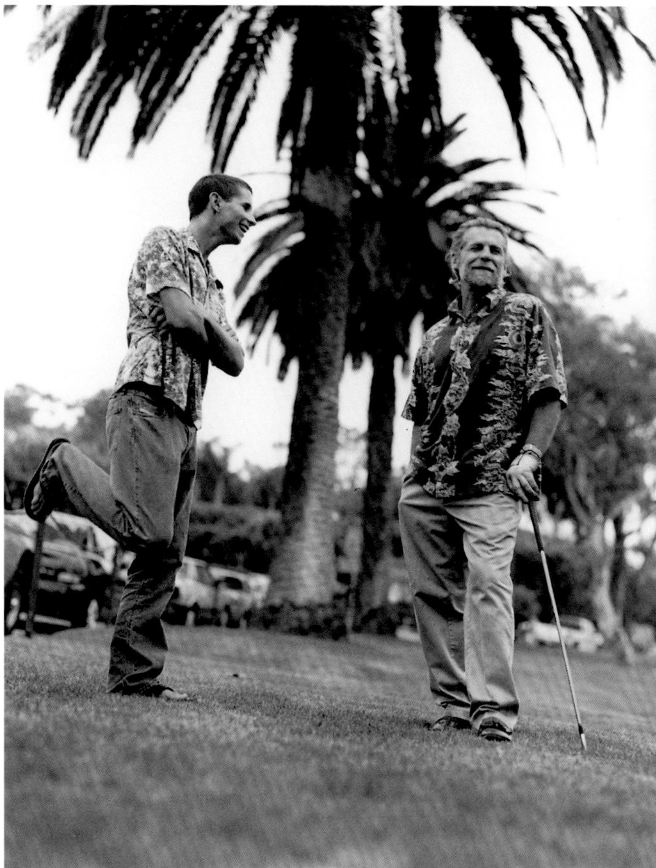

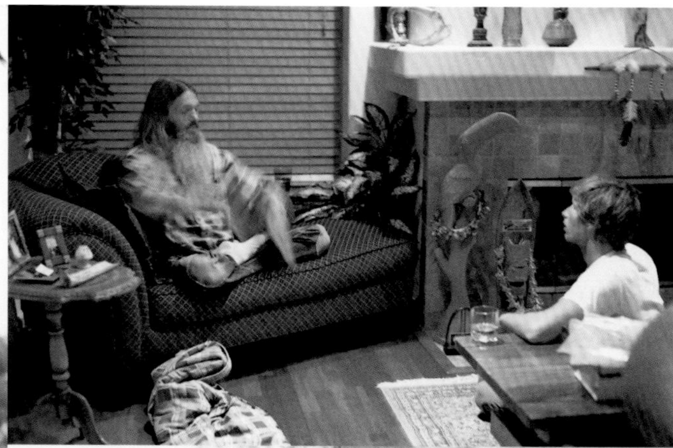

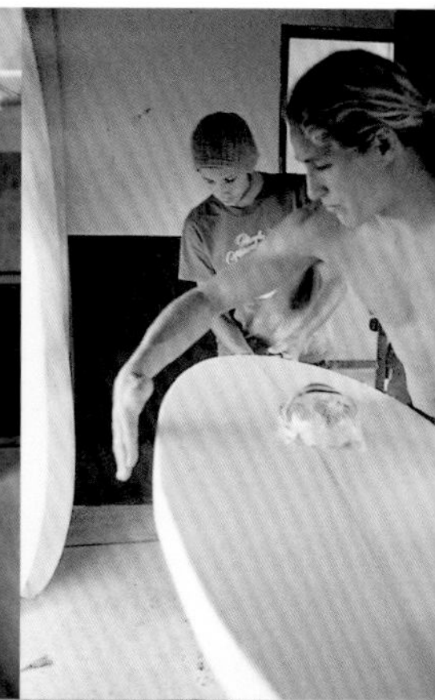

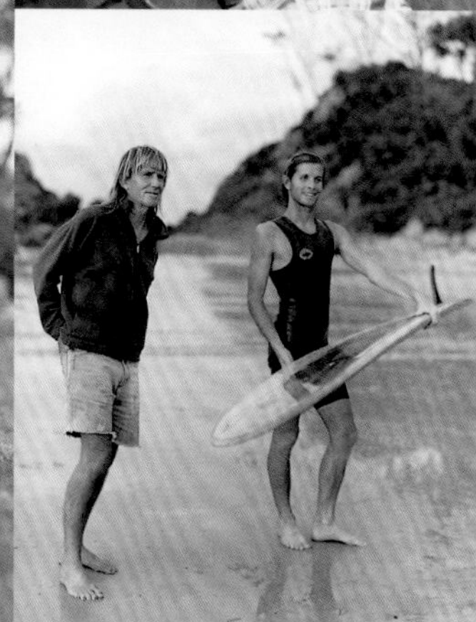

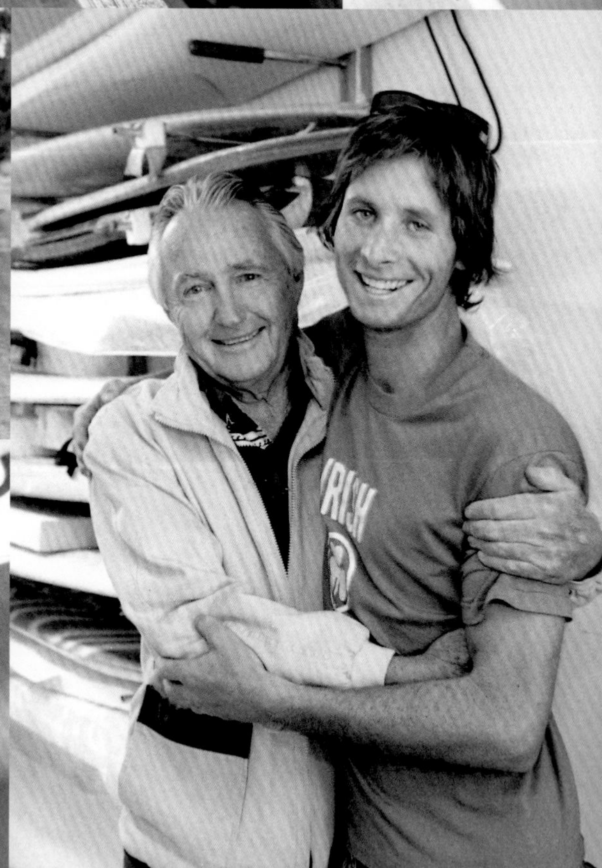

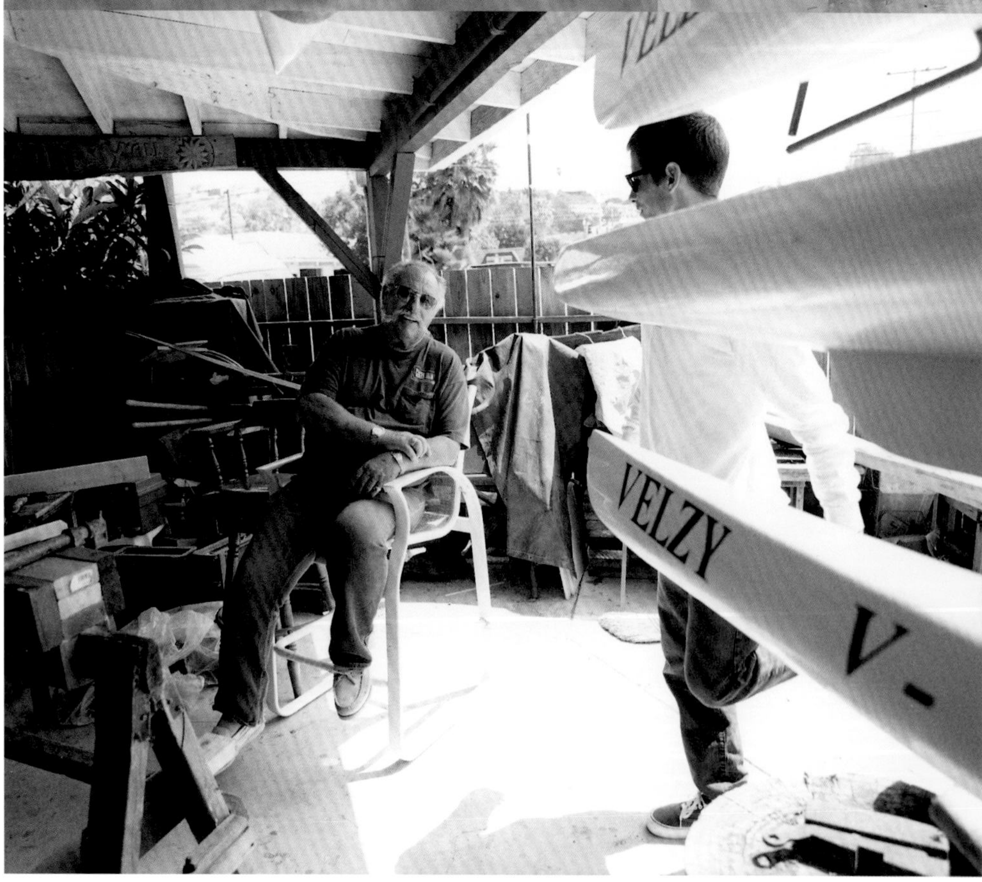
VELZY
V-

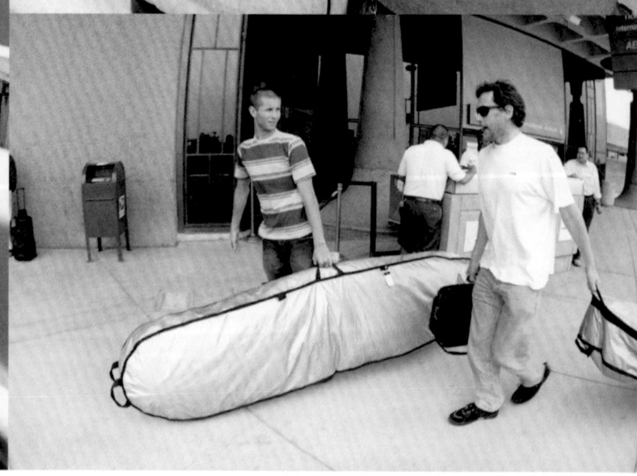

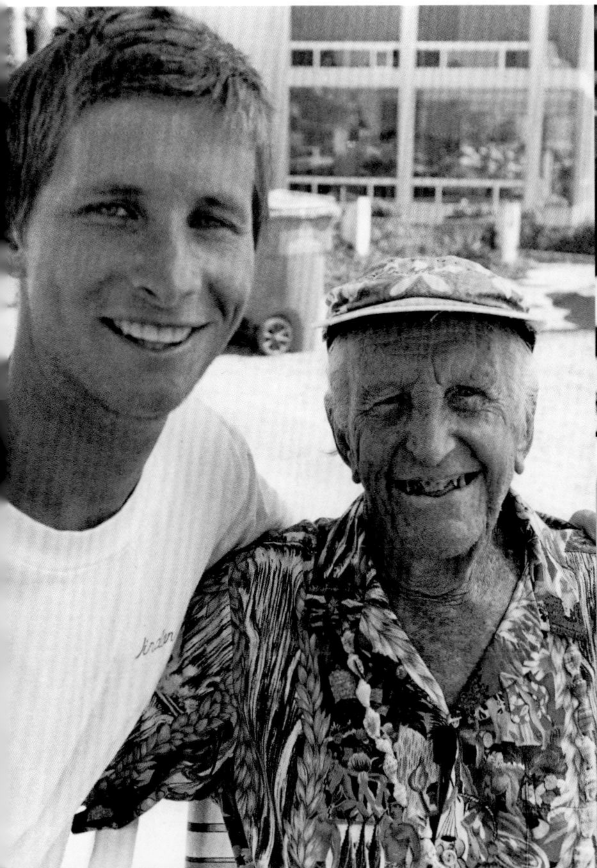

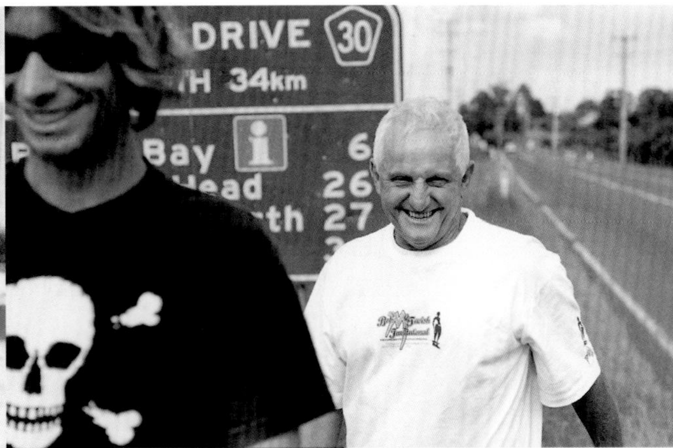

DRIVE
30
Bay 6
Head 26
27

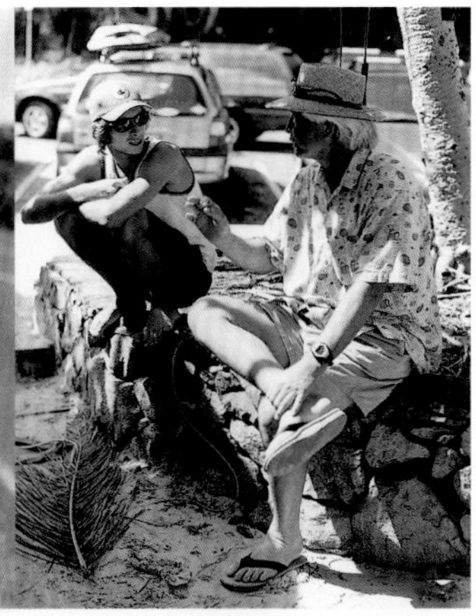

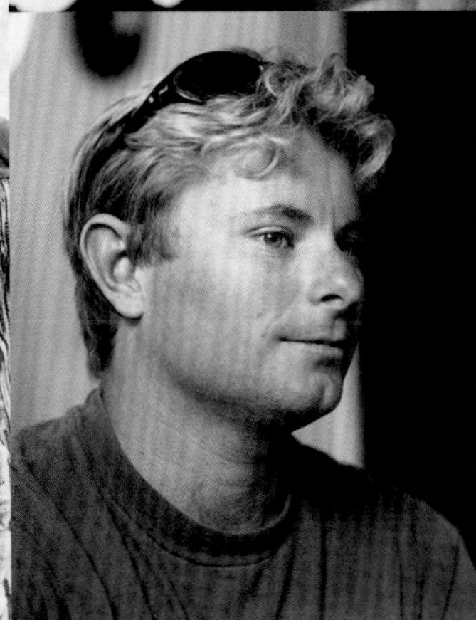

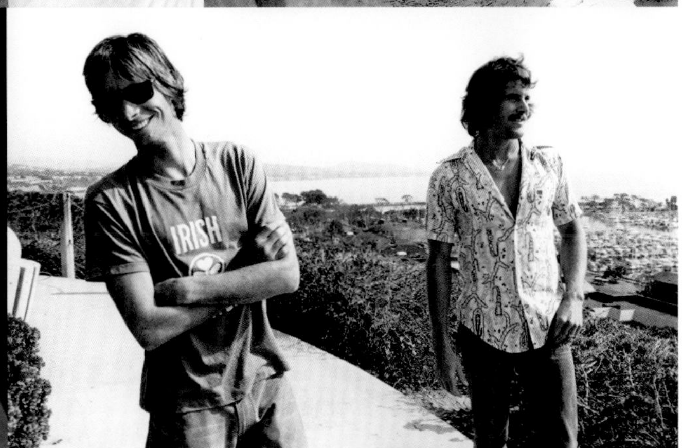

IRISH

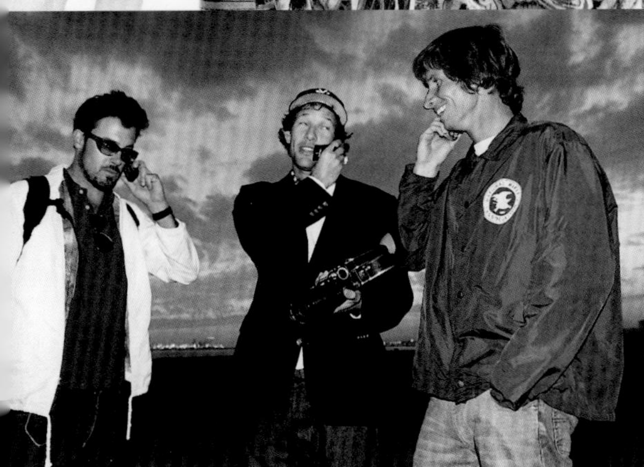

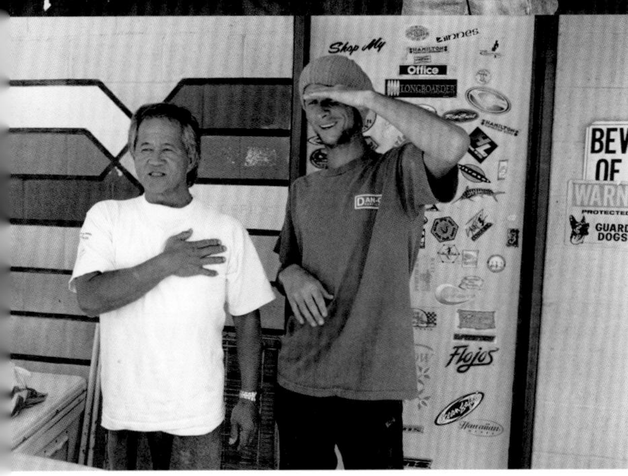

Office
LONGBOARDER
GUARD DOGS
Flojos

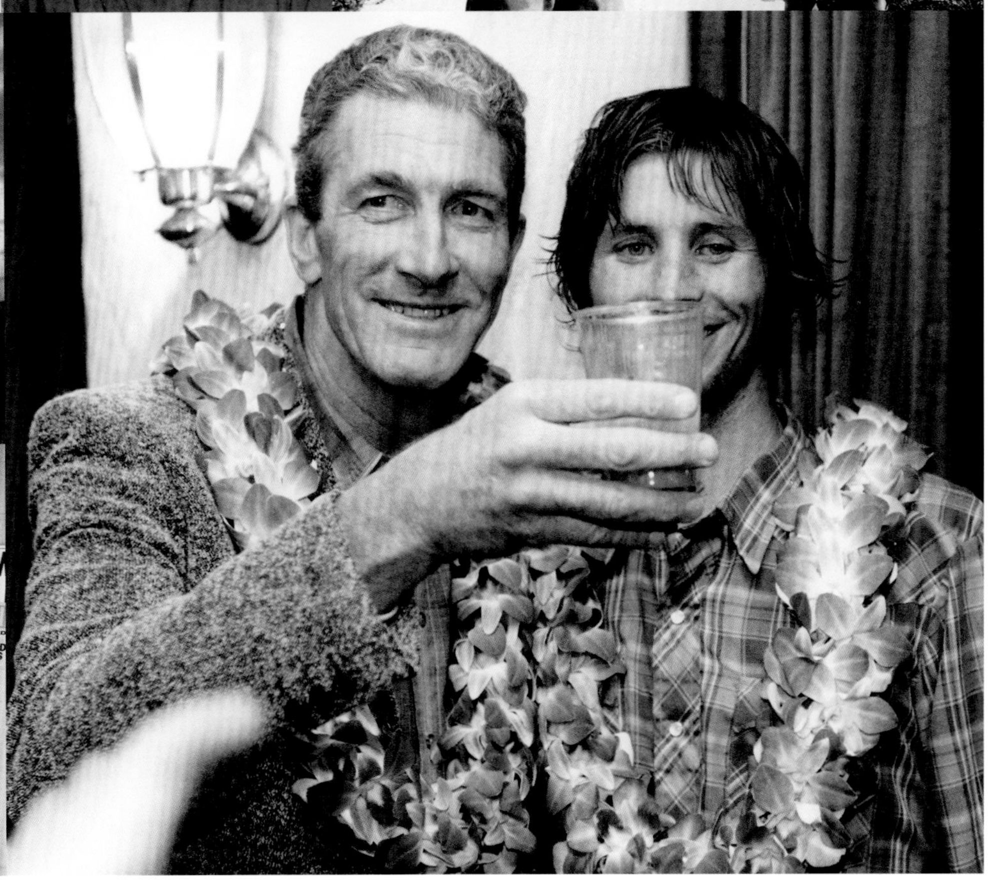

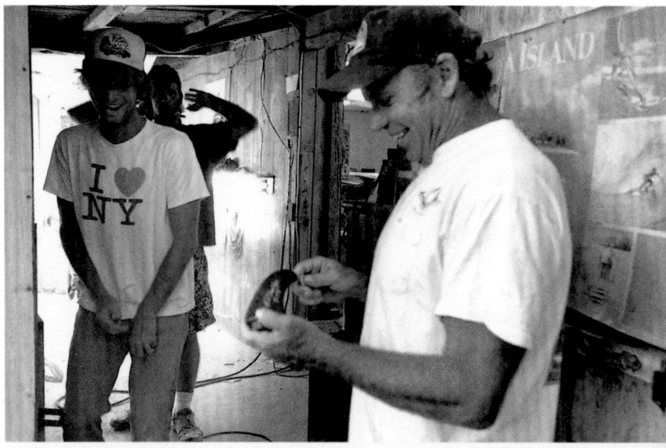
I ♥ NY

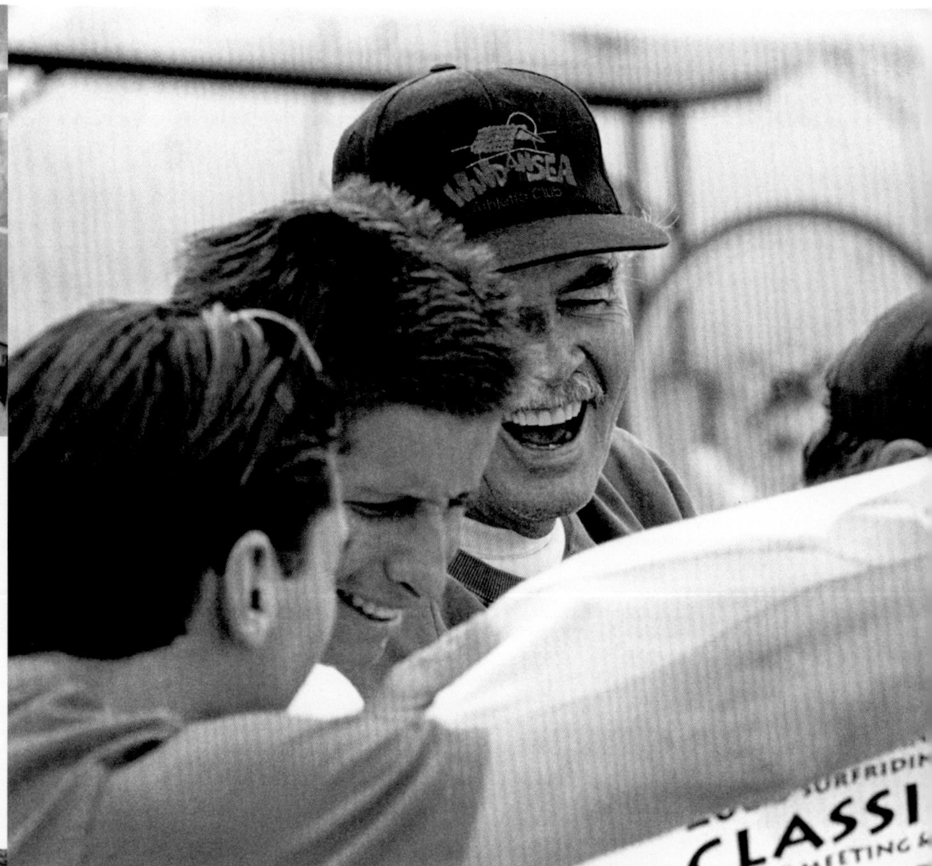
WINDANSEA

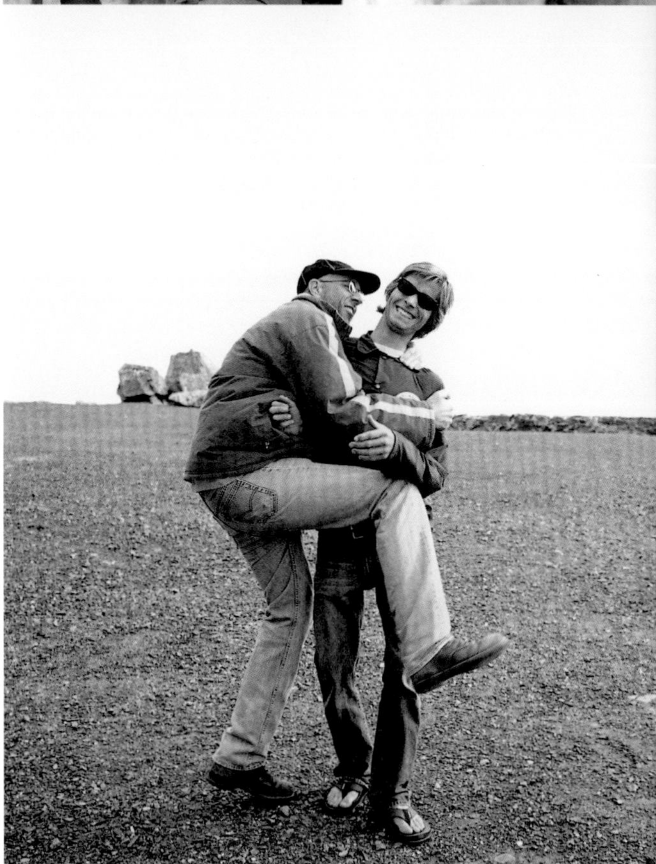

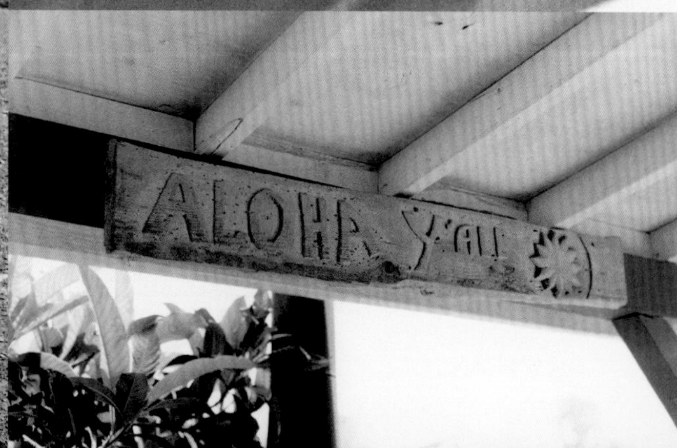
ALOHA Y'ALL

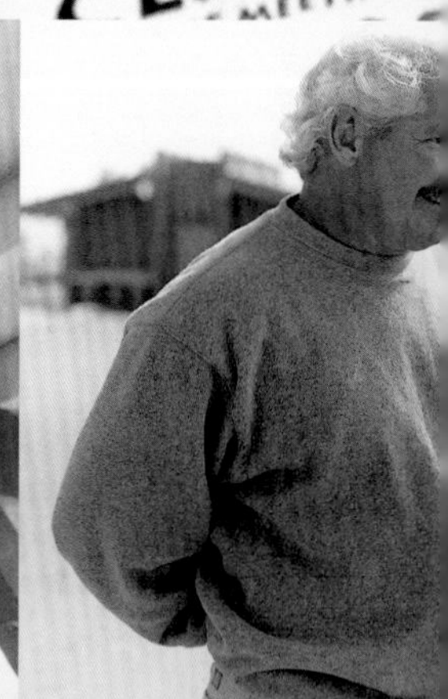

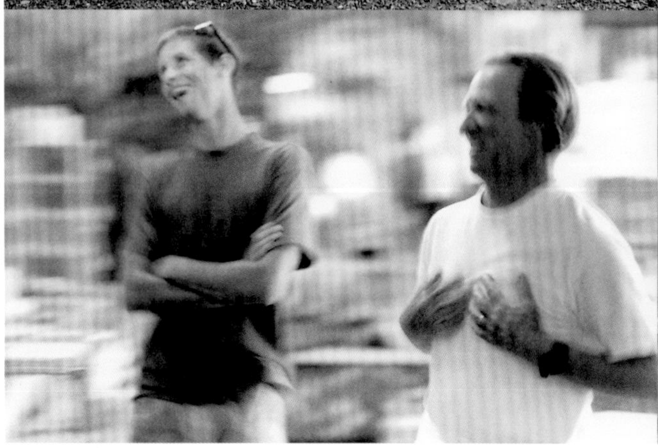

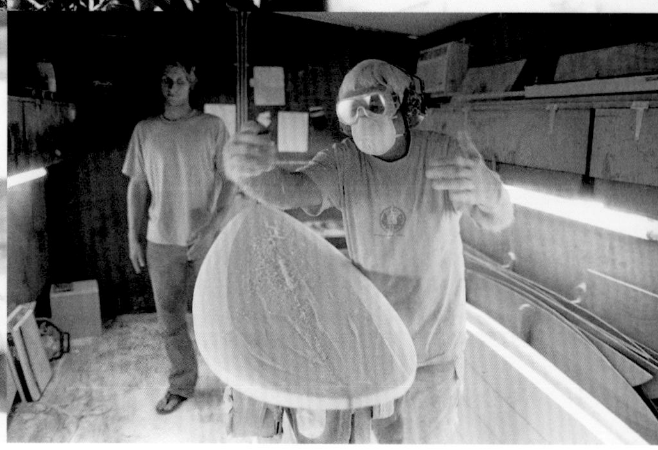

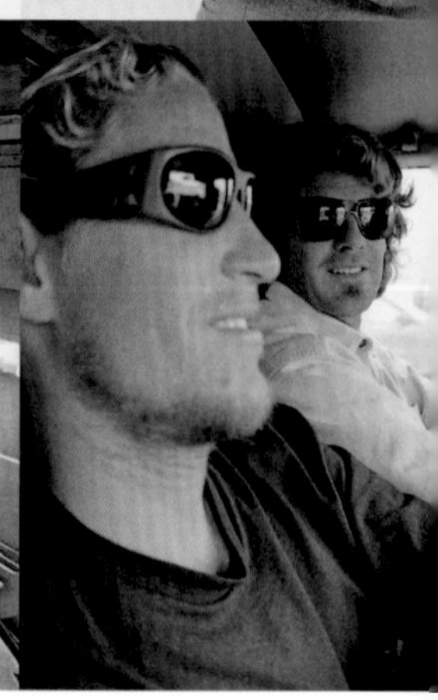

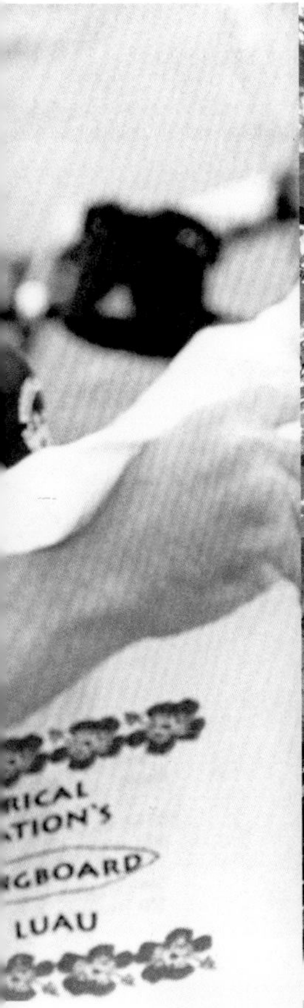
LUAU
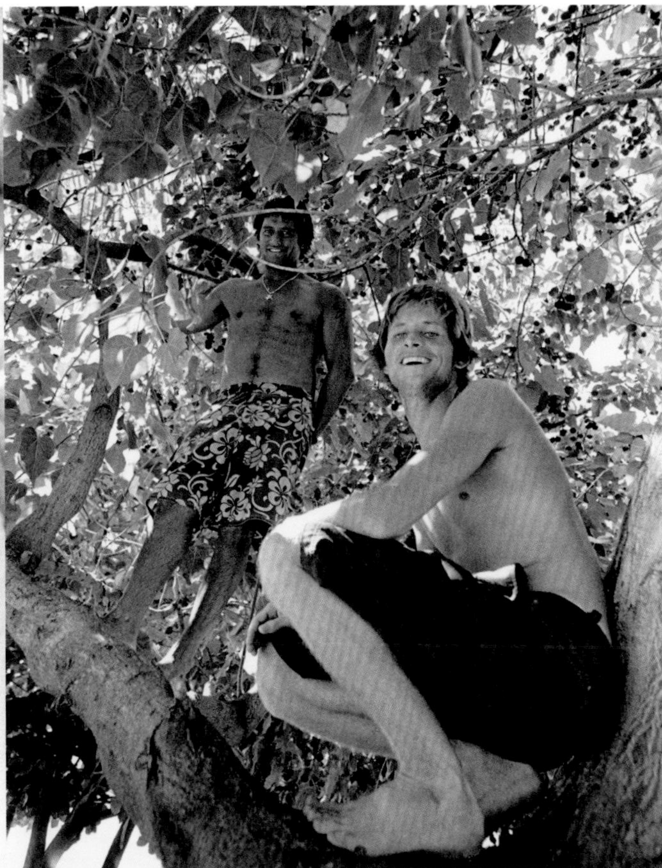
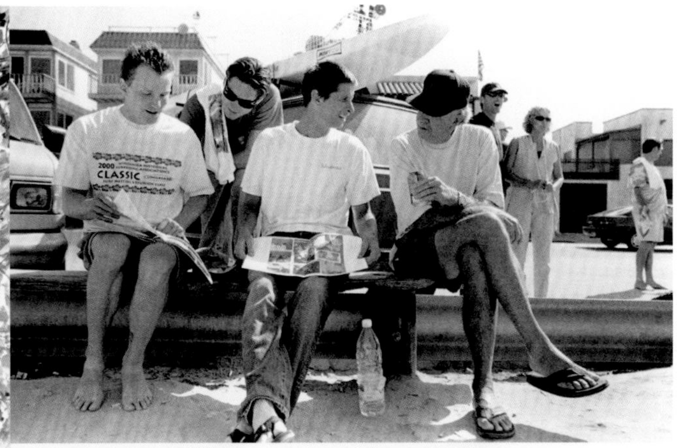
CLASSIC
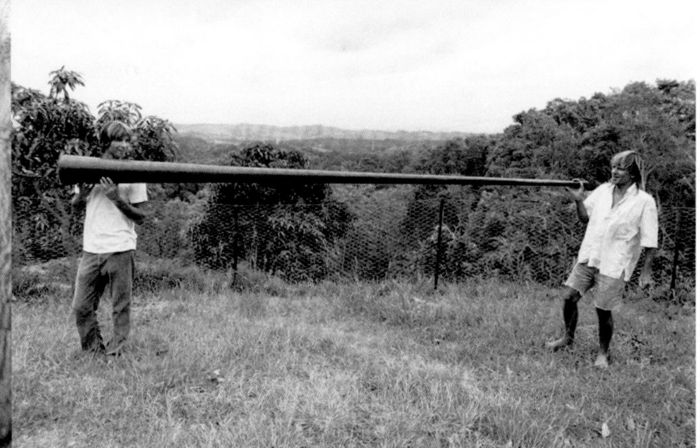
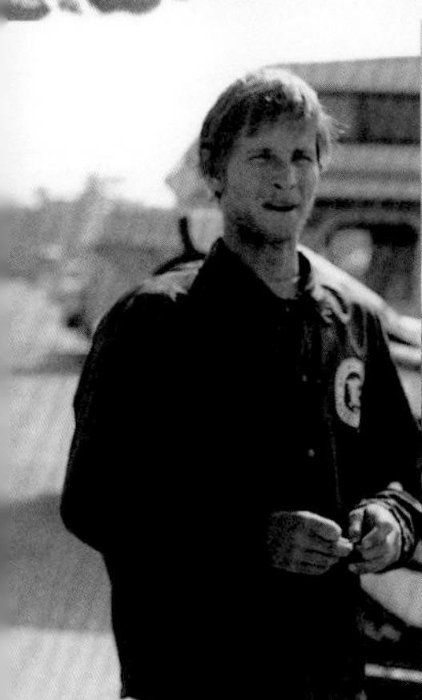
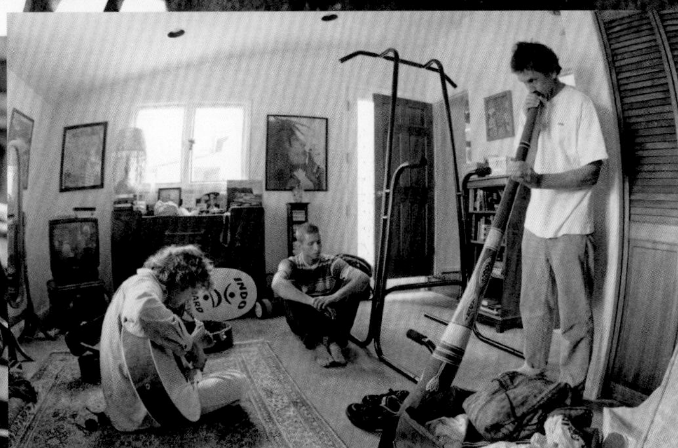
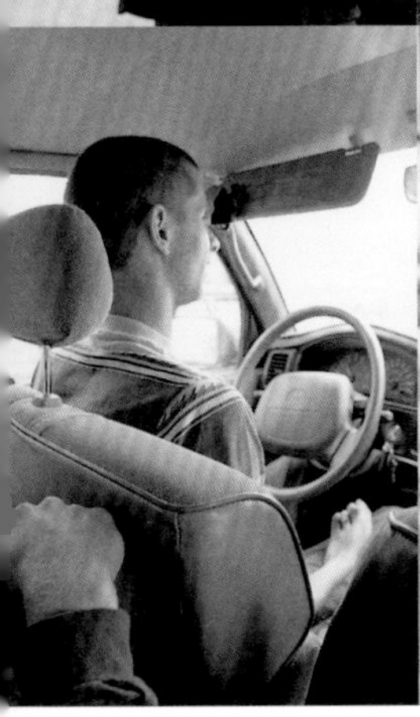
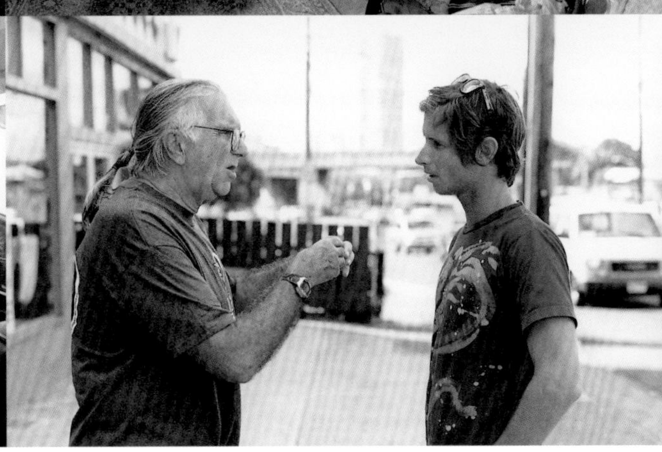
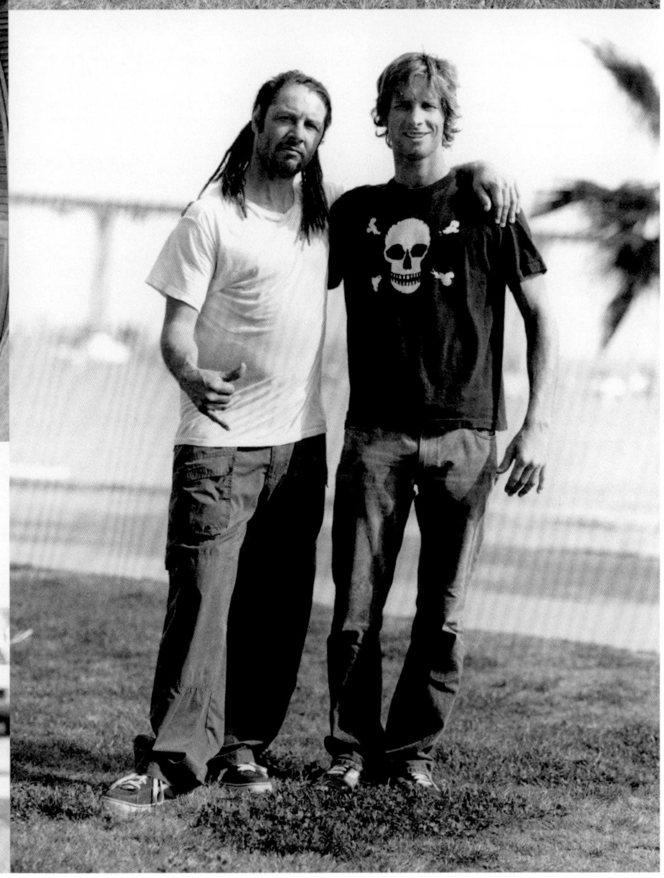

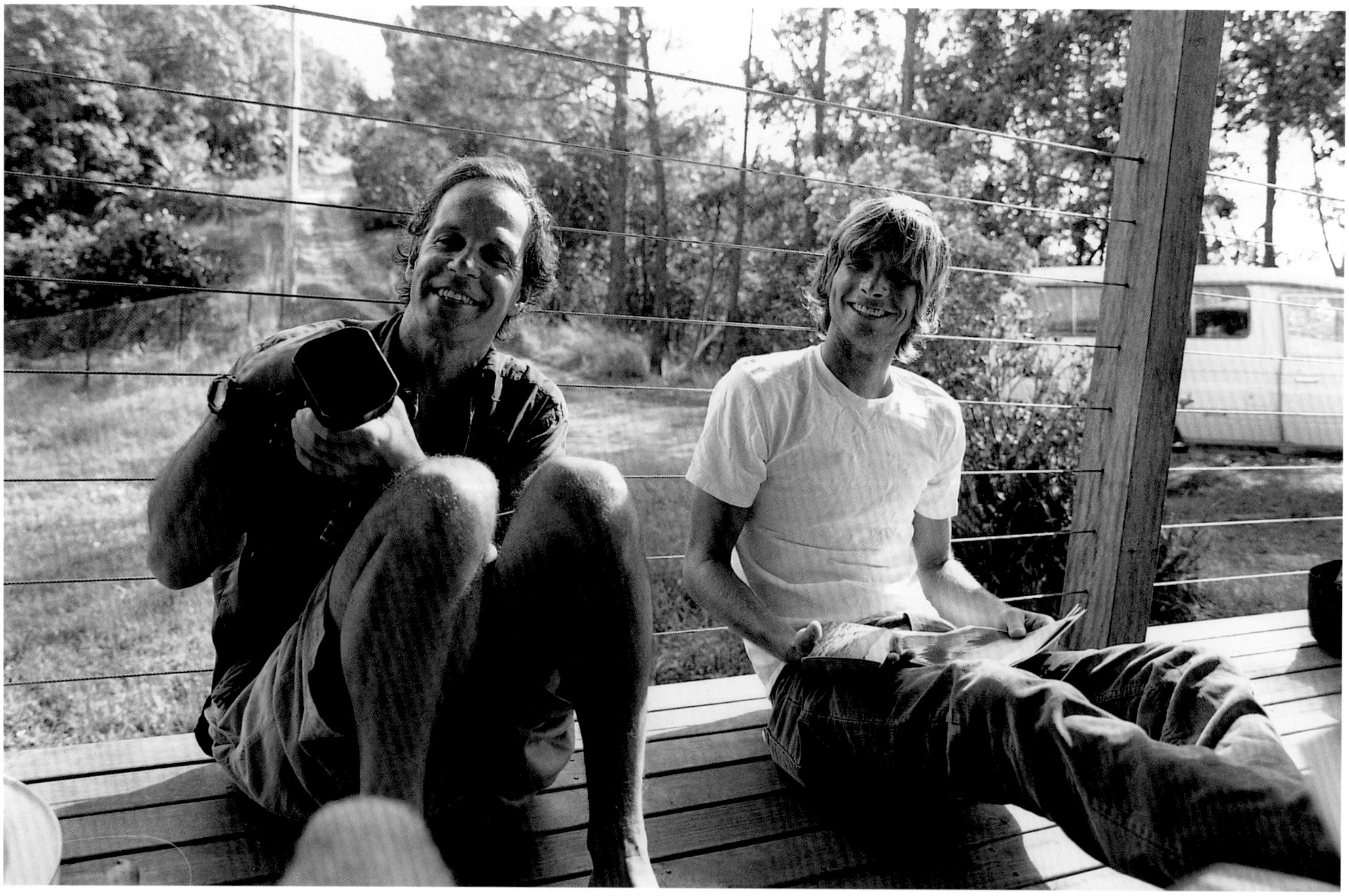

Photograph by George Greenough

Michael Halsband Acknowledgements

Dedicated to my mother Carole Rosenberg

Carole & Alex Rosenberg
I could not have done any of this without your unending love and support.
Donald Takayama
a great friend and teacher, and for putting me up and putting up with me.
Raja Sethuraman
for your patience, support, generosity, and friendship.
Jane Rose
my soul sister and great friend.

A very special thank you for all the love and support

Susan Anton, Bernie Baker, Bjorn De Boer, Bob Brooks, Bryan "Bubba" Charron, Pat & Clare Anne Conlon, Richard Fortus, Jerome Halsband, Kenneth Halsband, Dan Hopkins, Devon Howard, Sofie Baron Howard, Dr. Debra Jaliman, Buzzy Kerbox, Darren & Lisa Leddingham, Betsy Long-Phillips, Rudi & Wendy Manheim, Takuji Masuda, Louis William Max, John McWhinnie & Maria Beaulieu, Jane Miller, Keki Mingus, James Moore, Barry Neuman, Doc Paskowitz, Israel "Izzy" Paskowitz, Jonathan Paskowitz, Serene Pelletier/Da Kine, Bonga Perkins, Charley Randazzo, Lawrence Reed, MD., Larry Reed, Will & Carrie Regen, Matt Rimel, Elisabeth Robinson, Jim Russi, Walter Silver, Hershal Sklaroff, Mike, Claudia & Malik Solomon, Sydney Lauren Stutterheim, Rick Tamarian, Barrett Tester, David Tew, Aleric Valentine, and Chris Wontoloa

Studio

Lindsay Blatt, Michelle Ip, Austin Kennedy, Vanessa Lenz, and Jennifer McClure

Black and white printing by Lexington b+w

Laurent Girard, Jeffrey Kane, Peter Berberian, and Bibi Salim-Richards

Color photographic processing and printing by Color Edge Art

Chia Chen, Max Saravanan, Alex Mindin, Peter Giglio, John Lee, Frankie Delvalle, Jesus Mejia, Jasmin Hurtas, Jose Rivera, and Sammy Rossi

Color Edge Digital

Jacques Wolf, Sean Barry, and Norman Goben

Produced by Michael Halsband Studio
1200 Broadway, New York, NY 10001, USA
Phone 212 889 2994 Fax 212 683 7826
www.michaelhalsband.com

Produced in association with Channel Photographics LLC
116 East 16th Street, 12th Floor, New York, NY 10003, USA
Phone 212 254 5240 Fax 212 254 5246
www.channelphotographics.com

Distributed by SCB DISTRIBUTORS
15608 South New Century Drive, Gardena, CA 90248-2129, USA
Phone 310 532 9400 Fax 310 532 7001

Printed in China by Palace Press International

First edition, 2004

Library of Congress Control Number: 2004110218

ISBN: 0-9744029-4-X

Book Design: Douglas Lloyd and Andy Boonthong
Lloyd (+co), New York

Project Consultant: James Crump

Text Editing: John Yau and Peg Goldstein

Special Thanks: Michiaki Ishida and Amsterdam Wetsuits

A special edition of Surf Book is limited to fifty copies that are signed, numbered and are accompanied by a photographic print by Michael Halsband contained within a clamshell presentation box.

A signed and numbered slipcased edition of Surf Book is also available and limited to 100 copies.

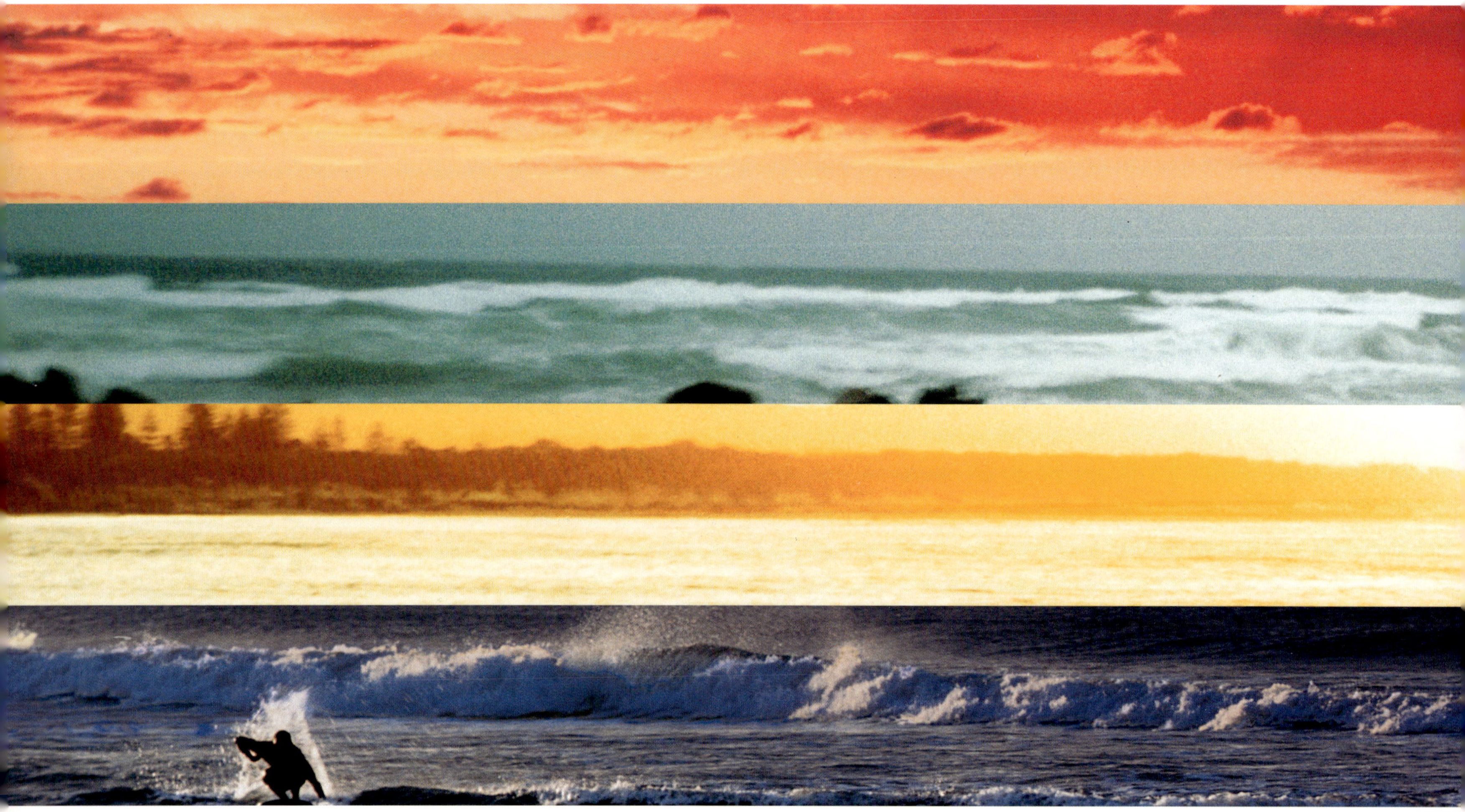